D0617244

Privatization of Information and Agricultural Industrialization

Privatization of Information and Agricultural Industrialization

Edited by
Steven A. Wolf

CRC Press
Boca Raton New York

Library of Congress Cataloging-in-Publication Data

Catalog information may be obtained from the Library of Congress

Direct all inquiries to CRC Press LLC, 2000 Corporate Blvd., N.W., Boca Raton, Florida 33431.

International Standard Book Number 1-57444-104-3
Printed in the United States of America 1 2 3 4 5 6 7 8 9 0
Printed on acid-free paper

TABLE OF CONTENTS

ABOUT THE CONTRIBUTORS

Michael Boehlje is a professor in the Department of Agricultural Economics at Purdue University where he is conducting research and teaching in the area of farm and agribusiness management and finance. His research interests include alternative systems of coordination of the food and industrial product chain, the industrialization of agriculture, and alternative financial and organizational structures for farm and agribusiness firms. Before moving to Indiana, Dr. Boehlje was a member of the faculty at the University of Minnesota and Iowa State University. Born on a farm in northern Iowa, he was educated at Iowa State and Purdue Universities.

J. Bruce Bullock is a professor, Department of Agricultural Economics, University of Missouri - Columbia. From 1989 to 1994 he was the associate director, Missouri Agricultural Experiment Station. Dr. Bullock received a B.S. and M.S. from Oklahoma State University and worked in the USDA Economic Research Service before earning his Ph.D. in agricultural economics from the University of California - Berkeley. Prior to moving to Missouri, he was a member of the faculty at North Carolina State University, Oklahoma State University, and a visiting assistant professor at the Universidad Agraria in Peru.

Michael Bunney is a partner in the Performance Improvement Partnership, a British business development consultancy, and a member of the Guild of Agricultural Journalists. Previously, he was marketing and market research manager for eight years at the British Agricultural Development Advisory Service. Much of his work, both within public agencies and as a private consultant, focuses on developing a more market-oriented and commercial approach within the agricultural industry. Mr. Bunney started his career as an agricultural advisor (field agent) following a degree from the University of Reading.

Lawrence Busch is a professor of Sociology at Michigan State University. He was born and raised on a very small farm in New York City. He received his Ph.D. from Cornell University. His research focuses on agricultural research, extension, and higher education policy, both in the United States and abroad. He recently co-authored *Making Nature, Shaping Culture: Plant Biodiversity in Global Context* (Nebraska 1995). He is currently beginning a volume on the dilemmas of development. He is a Fellow of the American Association for the Advancement of Science. He serves on the scientific advisory board for CIRAD, the French foreign agronomic research agency.

Frederick H. Buttel is a professor of Rural Sociology and Environmental Studies, and co-director of Agricultural Technology Studies at the University of Wisconsin, Madison. His major research interests lie in the social and environmental implications of agricultural research and technological change in agriculture. He was President of the Rural Sociological Society in 1990-91, and was elected a Fellow of the American Association for the Advancement of Science in 1987.

John Cary is an associate professor and head of the Department of Agriculture and Resource Management, University of Melbourne, Australia. He has been a visiting fellow at the University of Wisconsin and the University of Michigan, a visiting scientist at Landcare Research New Zealand Ltd., and a consultant. He is co-author of the book *Greening a Brown Land: The Australian Search for Sustainable Land Use*. His current research is focused on the psychology of landscape assessment, public perceptions of landscapes, and the promotion of biodiversity and ecotourism.

Marie-Claude Fortin is a research scientist with the Pacific Agriculture and Agri-Food Research Centre, Agriculture and Agri-Food Canada Research Branch, Cloverdale, British Columbia where she is part of a team specialized in the design of decision-support systems for agriculture. She has previous experience in conservation tillage and crop management. Dr. Fortin received her B.S. and M.S. from McGill University and her Ph.D. from Michigan State University.

Don Holt is an associate dean, research director, and professor of Crop Sciences in the College of Agricultural, Environmental and Consumer Sciences, University of Illinois. His undergraduate and masters degrees in Agronomy are from the

University of Illinois. After farming for seven years he returned to earn his Ph.D. from Purdue University where he went on to serve as a member of the Agronomy faculty for fifteen years. Dr. Holt is a fellow of the American Society of Agronomy, Crop Sciences Society of America, and American Association for the Advancement of Science. He served as president of the American Society of Agronomy and the Agricultural Research Institute.

Nicholas Kalaitzandonakes is an associate professor, Department of Agricultural Economics, University of Missouri - Columbia. Dr. Kalaitzandonakes received his B.S. from the College of Athens, Greece and his Ph.D. in Agricultural Economics from the University of Florida. His research interests include interaction between industrial organization and innovation processes, technology transfer, and measurement of productivity growth. In addition to teaching and research, he has served as a consultant to OECD and USAID.

Dennis Keeney is the director of the Leopold Center for Sustainable Agriculture since 1988 and has been the director of the Iowa State Water Resources Research Institute at Iowa State University since 1991. A native of Runnells, Iowa, he received his Ph.D. in Soil Science from Iowa State. He joined the faculty of the University of Wisconsin Department of Soil Science and went on to chair the Soils Department and the Land Resources Graduate Program of the Institute for Environmental Studies. Dr. Keeney has conducted research on a range of topics with specific emphasis on agronomic and environmental aspects of nitrogen management. He is past president of both the American Society of Agronomy and Soil Science of America.

Madhu Khanna is an assistant professor of Environmental Economics at the Dep artment of Agricultural and Consumer Economics, University of Illinois at Urbana-Champaign. She received her Ph.D. from the University of California - Berkeley Department of Agricultural and Resource Economics. The focus of her research has been policies to induce adoption of precision technologies, pollution prevention and sustainable economic growth. She is currently working on the impact of adoption of precision technologies on regulation of non-point pollution.

Gerad Middendorf is a doctoral candidate in the Department of Sociology at Michigan State University. He is interested in the social implications of technology development in international and domestic agriculture. His M.A. thesis in sociology is entitled, "*Inquiry for the Public Good: Citizen Participation in Agricultural Research.*" He also holds an M.A. in International Affairs from Ohio University. His interest in international development issues stems from work in agricultural extension with the Peace Corps in Honduras.

Francis J. Pierce is a professor of Soil Science in the Department of Crop and Soil Sciences at Michigan State University. He received his M.S. and Ph.D. degrees in Soil Science from the University of Minnesota in 1980 and 1984 and has been at MSU since 1984. His research interests have focused on soil management and soil and water conservation, with an emphasis on conservation tillage, soil compaction, erosion effects on soil productivity, and fate of nitrogen. He has focused recently on evaluation of site-specific management for agriculture and precision farming.

Amy Purvis is an assistant professor at Texas A&M University. Since 1994, she has held a joint appointment in the Department of Agricultural Economics and the Department of Rangeland Ecology and Management. In 1992 and 1993, she was a bystander watching adaptive management in the dairy cluster around Stephensville, Texas. Her Ph.D. is from the Department of Food and Resource Economics at the University of Florida (1993). Her M.S. is from the Department of Agricultural Economics at Michigan State University. From 1982-1986, she served as a Peace Corps aquaculture extension agent in Zaire.

Elizabeth Ransom is a student in the Ph.D. program in the Department of Sociology at Michigan State University. Her interests are in the social impacts of new technologies in international and national communities, with particular emphasis on women. Other research interests include race, ethnicity, and women in sports.

Brian Scarsbrick is the chief executive of Landcare Australia Limited, a non-profit institution under contract to the federal government of Australia to promote formation of Landcare groups and seek private and corporate funding for promotion of Landcare. Landcare is a community-based approach to sustainable resource management started in 1989 and now consisting of 2,500 Landcare groups representing over 30% of Australia's farmers. Before joining Landcare, Scarsbrick spent sixteen years as an agricultural extension agronomist and ten years as an agricultural administrator with the New South Wales Department of Agriculture.

David Sunding is a senior economist on the President's Council of Economic Advisors. He specializes in agricultural and natural resource policy, especially regarding agricultural chemical use, public lands policy, water resources and land use planning. Dr. Sunding was formerly associate director of the Center for Sustainable Resource Development at the University of California - Berkeley, where he also taught in the Department of Agricultural and Resource Economics and at the Boalt Hall School of Law.

William T. Vorley has first hand experience of the potential and constraints for achieving sustainable development within the private sector. In 1993, Vorley resigned from Ciba to take a position as visiting scientist at the Leopold Center for Sustainable Agriculture. Under grants from the Leopold Center, Ciba, and USEPA, he is currently coordinating a multidisciplinary network of analysts to chart options for pesticide manufacturers in sustainable agriculture. This work will be published as a book in early 1997. Bill has an academic background in agricultural ecology. He is a British citizen, and has lived and worked in the UK, Japan, Malaysia, Indonesia, and Switzerland.

Steven A. Wolf is a post-doctoral Fellow in the Department of Agricultural and Resource Economics at the University of California-Berkeley. In 1996, he received his Ph.D. from the Institute for Environmental Studies, Land Resources Program, at the University of Wisconsin-Madison. He holds an M.A. in Urban and Environmental Planning from the University of Virginia. His principal research interests lie in understanding and improving the environmental and social performance of production systems through integrated assessment of historical, institutional, economic, technological, and ecological factors.

David Zilberman is a professor and chair of the Department of Agricultural and Resource Economics, and director of the Center of Sustainable Resource Development, at the University of California at Berkeley. He received his Ph.D. in Berkeley in 1979 and, since then, has been on the faculty in Berkeley. His research interests include the economics of natural resources in agriculture, economics of technological change, decision making under uncertainty, and environmental and agricultural policies. Dr. Zilberman has been an advisor to the USEPA, California State Department of Food and Agriculture, USDA, and international organizations.

Preface

This book is a product of a workshop held at the University of Wisconsin - Madison on October 25-26, 1995. That meeting, Privatization of Information and Technology Transfer in U.S. Agriculture: Research and Policy Implications, was attended by roughly ninety people. Attendees represented scholars from a variety of social and agricultural production science disciplines, small and large agribusinesses, state and federal government agencies, Extension, and advocacy organizations. The meeting was designed to enhance understanding and raise the salience of shifting patterns of public and private sector relations in development and dissemination of agricultural data, information, and technology.

This book was developed to stimulate research and debate. Our objective is to see questions associated with agricultural information and the social and institutional relations of information addressed with increasing frequency and sophistication in research publications, professional meetings, classrooms, and policy and planning forums. While this book is focused squarely on agricultural systems, questions concerning the philosophy and mechanics of governmental intervention in production systems have applicability to all spheres of economic activity.

The papers contained in this volume represent presentations made during the October 1995 workshop, as well as development of authors' thinking after returning from the meeting. Three principle themes hold these papers together as a collection. First, there is common acknowledgment that institutional relations governing development, control, and application of information in agriculture are currently changing at an accelerated pace. Second, there is agreement among authors that understanding of historic, current and future processes through which knowledge is created and information is applied in agricultural production systems will be developed through synthesis of political, economic, institutional, and technological considerations. The authors clearly recognize the interdisciplinary nature of these questions. Third, there is agreement that diminished public sector involvement and increased private sector responsibility for information development and dissemination in agriculture is significant. Privatization of information raises meaningful social, economic, and environmental concerns deserving of immediate attention by analysts, advocates, policy makers, and those with a direct economic stake in agriculture.

Organized in connection with my doctoral thesis in the Land Resources Program of the University of Wisconsin Institute for Environmental Studies, many people deserve acknowledgment for providing me with guidance, assisting in organizing, funding, and hosting the workshop, and preparing this volume. Peter Nowak, my colleague and thesis advisor, is deserving of special recognition for providing long-standing financial and intellectual support; Bob McCallister, for logistical, editorial, and personal support; members of the Program Steering Committee—Sandra Batie, Fred Buttel, Dennis Keeney, Peter Nowak, Edwin Rajotte, and David Zilberman—for valued leadership; James Bonnen, Roberta Parry, Stephen Harsh, and David Schweikhardt and others for their participation in a workshop planning session hosted by Sandra Batie at Michigan State University; Max Schnepf and Heather Hartley for consistently good advice; Stuart Eimer, Kevin Shelley, Brian Obach, Thomas Green, and Lee Gottschalk and the staff of the UW-CALS Conference Office for their efforts; Roger Wyse, Rick Klemme, Harold Lambert, Thomas Stein, Steve Watts, Grant Mangold, Kermit Burnside, Ray Hoefer, Peter Kuch, Katherine Smith, and Charles Rock for their presentations during the workshop; the American Journal of Agricultural Economics for permission to reprint Steven Wolf and Frederick Buttel's paper (Chapter 7); the authors of the individual chapters in this book for their contributions; and all of the people who attended the October 1995 workshop.

Conference sponsors provided the financial resources to convene the workshop and publish this volume, and I am grateful for their generous support. Sponsors included:

- U.S. Department of Agriculture, Cooperative State Research, Education, and Extension Service (National Research Initiative, Agricultural Systems award #9503199)
- U.S. Department of Agriculture, Natural Resources Conservation Service
- Ciba Plant Protection and Charles Rock and David Flackne
- The Farm Foundation and Walter Armbruster
- University of Wisconsin, Agricultural Technology and Family Farm Institute
- University of Wisconsin, College of Agricultural and Life Sciences, School of Natural Resources and Don Field
- University of Wisconsin, Institute for Environmental Studies

- American Society of Agronomy and Robert Barnes and Tom Hall
- Leopold Center for Sustainable Agriculture and Dennis Keeney
- Dow Elanco and Larry Wee
- U.S. Environmental Protection Agency and Peter Kuch and Roberta Parry
- The Soil and Water Conservation Society

The diversity in the types of organizations that chose to sponsor this project reflects shared understanding of the importance of research and discussion focused on these topics. Additionally, diversity in sponsorship reflects recognition of economic and political interdependence among a broad array of public and private institutions engaged in agricultural production, research, extension, education, and policy.

Steven A. Wolf
Berkeley, California

Introduction

Steven A. Wolf

The roles of public sector institutions and private sector institutions engaged in development and dissemination of information in agriculture are currently undergoing dramatic change. After more than one hundred years of public sector leadership in agricultural research and extension in the U.S., we appear to be entering a new era in which private and corporate interests have significantly increased authority and responsibility as information providers (ERS 1995; NAICC 1994; NRC 1996; Rivera and Gustafson 1991; Wolf 1995, 1996). The changing pattern of institutional relations raises questions concerning the social, economic, and environmental performance of agricultural systems.

This book describes and analyzes a variety of arrangements that contribute to the current mosaic of reorganization. In addition to new patterns of public-private relations in agricultural extension, private sector agents (e.g., producers, input industries, marketing cooperatives) are increasingly doing their own agricultural research and servicing their own information needs. Internalization of information by firms and the increasingly proprietary character of innovations in agriculture raise important questions. As Bonnen (1983) warned more than ten years ago, a situation in which public investments and political support for public sector engagement in research and extension are shrinking, while private information resources and incentives to make such investments are expanding, poses grave risks. In an era in which information is regarded as a crucial input in production systems, the opposite nature of public and private trajectories causes us to take notice.

To frame the questions addressed in this volume, I note that the primary focus is production, distribution, and application of production-oriented agricultural information in U.S. agriculture. However, our treatment does encompass other types of information, namely economic information (i.e., investment, pricing, and marketing), and developments in other countries. As an additional point of reference, our concern in this book is distinct from an analysis of institutional relations of agricultural research (ERS 1995; Huffman and Evenson 1993; NRC 1995; Ruttan 1982; Weaver 1993). However, institutional relations in research are obviously linked to those of information and are, therefore, embedded in our analysis (see Boehlje, Chapter 2 in this volume for a definition of information and a discussion of the relationship between research, knowledge, data, and information). Last, our effort is not to present an institutional analysis of Cooperative Extension at federal, state, or county levels, although processes of extension are of central importance (Bennett 1995). Our objective is to:

- Elucidate economic processes associated with agricultural information
- Describe historic and emerging structures and patterns of institutional relations in agriculture
- Outline a set of economic, social, and environmental concerns that arise through analysis of dynamics of public-private relations specific to agricultural information.

The transition from a publicly dominated sphere of activity to one governed by markets, contractual arrangements, and property rights is referred to in this volume as a variant of privatization. In the U.S. context, unlike the United Kingdom or New Zealand cases featured in this volume, privatization of institutions organized to provide agricultural information is not federal policy. Our use of the term privatization in this volume refers to a spatially and structurally diverse pattern of institutional transitions and changes in social relations of production, rather than an explicit program of federal reform. Considering previously articulated concerns about the future of agricultural Extension (e.g., NRC 1989:14) and a continuing wave of downsizing, restructuring, and initiation of fee-based services in various state Extension Services, this analysis would appear to have domestic salience. Language in the 1996 Farm Bill mandating the Secretary of Agriculture to authorize private sector consultants to write farm and ranch resource management plans associated with cost-sharing or compliance programs is a concrete illustration of current trends (Hall 1996). The timing of this analysis is consistent with international developments as well. Federally orchestrated privatization of agricultural extension in other countries during the 1980s and into the 1990s (Bunney, Chapter 12, and Cary, Chapter 10, in this volume; Rivera and Gustafson 1991) indicates that developments in the U.S. are related to an international context. Further, a generalized, global trend towards national politics of anti-interventionism—deregulation, liberalization, and privatization—indicate developments in agriculture are related to larger patterns of political and economic change (M. Watts 1994).

Whether viewed as a process of public divestiture consistent with fiscal austerity and demographic shifts associated with modernization in developed nations, an outgrowth of internationalization, marginalization of the nation-state, and an emerging policy of liberalization (Bonanno et al. 1994; McMichael 1994), displacement of public institutions by vertically integrated, vertically coordinated industries who see information self-sufficiency as key to capital accumulation, or a product of technological innovation and industrial restructuring (Schertz and Daft 1994), privatization of information is a significant event in the history of agricultural development. As we head into the twenty-first century, a century in which we will confront staggering population growth, ecological pressures, and complex competition for rural resources, the rationale for and means through which the public sector will participate in agriculture has become unclear, leading us to question the relationship between agriculture and society. Who will provide information to farmers who cannot afford private information services or who are disadvantaged due to geographically uneven market development? Through what institutions can we overcome market failures that result in underinvestment in projects with diffuse benefit streams and long-term payoffs such as basic research and environmental protection? What is the future of the USDA and land grant university system? And the larger question, through what processes other than market relations will agricultural development be conditioned to reflect social values?

This book is concerned with specifying the dimensions of the privatization of information process (Which types of information? Which commodity sectors? At what pace?), discussing social and economic causes and mechanisms (Why now? Through what processes? Why in the observed forms?), assessing the implications of these changes (What will be the impacts on structure, industrial organization, social relations, natural resources, long-term competitiveness?), and outlining possible responses from researchers, agribusinesses, and policy institutions.

While it is tempting to say that private firms are now more important than public institutions in terms of information flows, such a statement masks the heterogeneity and inter-sectoral interdependence that characterizes agricultural knowledge and information networks (ERS 1995; Holt in this volume). Interdependence is apparent when we consider that information provided to farmers and agribusiness by private sector consultants very often originates in public sector institutions (e.g., land grant universities, USDA). And, not unimportantly, public sector institutions are dependent on direct financial and less direct political support from segments of the agricultural community. State relations in agricultural information provision are obviously complex and, in many cases, distinction between the public and private sectors is difficult. Cooperative Extension, notably in its early relationship with the railroads and later the American Farm Bureau, and in the important role of citizenry committees in directing local extension, is an important example of integration between the public and private sectors (McConnell 1969). A second example worth noting is the Certified Crop Advisor (CCA) program, a recently initiated, voluntary certification program proving to be very popular among private sector agronomic consultants. The USDA provided funding and served as a catalyst in the formation of the CCA program (S. Watts 1996). Lastly, university-based offices of technology transfer represent a key illustration of public and private sector interaction in development and marketing of agricultural innovations (Postelwaite et al. 1993). While public institutions remain deeply involved, their status is diminished as they move into supporting roles.

This collection of papers addresses restructuring processes from a perspective that tends to emphasize generalized trends and large-scale transitions in agriculture, rather than local adaptation and heterogeneity. The theoretical limitations and empirical contradictions that arise from an "unilinear, homogenizing, deterministic" approach to scholarship are increasingly recognized (Buttel 1994; Goodman and Watts 1994; van der Ploeg 1993; Wells 1996). While many of the chapters include case studies or references to specific community-level patterns of development, this book would have benefited from more intensive treatment of local models of diversity in production and dissemination of information. Having offered this self-directed criticism, I note that the chapter authors are aware of current data and knowledge limitations and are committed to further research.

The arguments made by authors in this volume indicate that we are in the midst of a transition in institutional relations regulating agricultural information. At the same time, it is clear that we are just beginning the process of documenting, through empirical assessment, the social and economic structures which define current and emerging information networks. As is often the case in dynamic systems, the theoretical frontier is in advance of the empirical frontier, and while this arrangement does not cripple our ability to describe developments with a measure of confidence, it does lead us toward a posture of humility. This book seeks to raise issues for discussion, contribute to theory development, and catalyze empirical research.

This book is not a strategic attempt to retain university-based agriculture researchers' piece of the shrinking public pie. Further, I do not argue that the historic significance of public institutions in U.S. agriculture (Huffman and Evenson 1993) is grounds for rubber stamping appropriations, nor that the excesses of capital disqualify industry for leadership positions. The objective here is to raise issues and re-emphasize the value of thorough exploration of the ramifications of political choices. I argue for conservatism in the current environment, an environment in which the public sector is often characterized as bloated, inept, and corrupt, while free markets and allocation of property rights are often viewed as universal prescriptions. Institutional innovation in agriculture, a sector fundamentally essential to our survival and quality of life, must be based on research, analysis, and debate rather than current fashion.

ORGANIZATION OF CHAPTERS

The first section, *Status and Trends in Agricultural Information*, is devoted to a discussion of the basic concepts confronted in an analysis of privatization of agricultural information. This initial set of papers also identifies opportunities for market failure and related risks that accompany movement towards privatization. In Chapter 1, Steven Wolf provides a general discussion of the context in which information demands in agriculture are changing and identifies a number of problematic implications of diminished public sector engagement in agricultural information development and dissemination. In Chapter 2, Michael Boehlje reviews a variety of relevant economic concepts and provides a definition of information that allows us to see its relationship to research, knowledge, and data. Boehlje's analysis succeeds in linking changes in the organization of agro-food systems to changes in patterns of demand for information. In Chapter 3, Dennis Keeney and William Vorley address the shortcomings of a system of information provision based on market discipline rather than principles of sustainability. Relying on the efficiency-substitution-redesign framework of Macrae et al. (1993), Keeney and Vorley argue that privatization, in the absence of additional institutional innovations designed to avert market failures, is inconsistent with the principles of sustainable development and sustainable agriculture. In Chapter 4, Don Holt discusses the complex and complementary relationship between the private sector and public institutions in conducting agricultural research and extension. Holt's analysis suggests that privatization is consistent with private firms' interest in accelerating and targeting the research and development process in an increasingly competitive economic environment.

The second group of papers, *Organizational and Technological Change in Production Systems*, addresses the production and control of information and technological inputs in farming systems. The papers describe institutional and technological changes impacting the nature and quality of information available to producers and the means through which information is produced and distributed. In Chapter 5, Elizabeth Ransom, Lawrence Busch, and Gerad Middendorf describe the impact of biotechnology on the status of agricultural cooperatives and identify a range of strategies cooperatives are employing to participate in increasingly capital-intensive research and development. Analysis of the future of cooperatives is significant in that cooperatives represent an institutional structure through which producers can conceivably pool resources in order to retain a measure of control over information production, products of biotechnology, and other inputs. In Chapter 6, Marie-Claude Fortin and Francis Pierce describe the principle components and theory underlying a computerized agricultural information system. Fortin and Pierce argue that a comprehensive, digital, spatially referenced database is consistent with the increasingly complex character of farm-level information requirements. In Chapter 7, Steven Wolf and Frederick Buttel address the significance of emergence of precision farming, a system of satellite-aided data recording and navigational control. Wolf and Buttel find that consistency between precision farming and a range of political and organizational challenges confronting agricultural industries helps to explain a great deal of the current enthusiasm surrounding precision technologies.

Chapters 8 and 9 address farm-level production environments directly. In Chapter 8, Amy Purvis reviews livestock production systems with particular focus on institutional relations in development of information used to address a range of natural resource conflicts. Based on analysis of industrialization in the cattle feeding, poultry, swine, and dairy sectors, and attendant implications of structural, organizational and technological change in each, Purvis identifies a set of relevant research questions that support community-based dispute resolution and a policy of adaptive management. In Chapter 9, Steven Wolf reviews the emerging structure and dynamics of private sector agrichemical management service markets. Based on analysis of national institutions and community-level data from states in the Midwest, Mississippi Delta, and Chesapeake Bay region, Wolf identifies the shifting roles of agrichemical dealers, independent crop consultants, and Extension in shaping fertilizer and pesticide management practices.

The third section of the book, *International Case Studies*, describes and analyzes privatization initiatives in the United Kingdom, New Zealand, and Australia. It is important to note in considering what is known about the performance of market-based approaches to development and dissemination of agricultural information that no systematic assessments or impact studies have been conducted to date. Based on what limited data are available and personal involvement of the authors, these papers provide valuable insight into processes of adjustment to a "user pays" system. These papers describe a range of institutional models that stand in contrast to those observed in the U.S. In Chapter 10, John Cary provides a conceptual and historical framework in which to assess variants of privatization of agricultural extension services. Cary describes and compares the experience of

New Zealand, which sold off its extension service, and developments in several Australian states. In Chapter 11, Brian Scarsbrick describes the Landcare movement in Australia. Landcare represents a community-based approach to agricultural extension, natural resource management, and ecological restoration. Scarsbrick's treatment focuses on the political and ecological context in which Landcare has met with success, and the role of corporate sponsorship in promoting the program. In Chapter 12, Michael Bunney provides a detailed description of historical developments and institutional processes of adjustment associated with a transition from publicly provided agricultural extension to a system of fee-based advisory services in the United Kingdom. Bunney's analysis highlights the phases through which privatization is proceeding and the respective roles of public and private institutions at each stage.

The fourth and final section of the book addresses *Institutional Innovation*. These two papers address the critical challenges and avenues for reform confronting the land grant university system. In Chapter 13, Nicholas Kalaitzandonakes and J. Bruce Bullock identify a critical mismatch between the information requirements of contemporary agriculture and the organizational behavior of land grants and their personnel. Their analysis of transaction costs indicates that university researchers are currently poorly positioned to provide the types of information and services required from public institutions, specifically user-friendly packages of decision-focused, multidisciplinary information. Kalaitzandonakes and Bullock suggest that without significant reform of the professional reward system and institutional culture, land grant universities will be increasingly disconnected from farmers, agribusinesses, policy makers, and other key clientele.

In Chapter 14, David Zilberman, David Sunding, and Madhu Khanna address the implications of organizational, institutional, and technological changes for Extension. Relying on economic principles and research findings, these authors identify continued and enhanced opportunities for Extension to provide leadership. Zilberman and co-authors identify consistencies between the knowledge and information requirements of twenty-first century agriculture, societal demands on agriculture, and the strengths of Extension.

REFERENCES

Bennett, C.F. 1995. Rationale for public funding of sustainable agriculture extension programs. Paper presented at American Evaluation Association Annual Meeting. Vancouver, B.C., Canada. November 1-5.

Bonanno, A., L. Busch, W. Friedland, L. Gouveia, and E. Mingione, (eds). 1994. From Columbus to Conagra: The Globalization of Agriculture and Food. Lawrence: University Press of Kansas.

Bonnen, J. 1983. Historical sources of U.S. agricultural productivity: Implications for R&D policy and social science research. American Journal of Agricultural Economics, Dec. 1983: 959-966.

Buttel, F.H. 1994. Agricultural change, rural society and the state in the late twentieth century: Some theoretical observations. In Agricultural Restructuring and Rural Change in Europe, D. Symes and A.J. Jansen (eds.). The Netherlands: Wageningen Agricultural University.

Cary, J. 1993. Changing foundations for government support of agricultural extension in economically developed countries. Rural Sociology XXXIII (3/4), pp. 336-347.

ERS (USDA/Economic Research Service). 1995. Agricultural research and development: Public and private investments under alternative markets and institutions. Staff Paper 9517. Washington, DC: USDA.

Goodman, D., and M. Watts. 1994. Reconfiguring the rural or fording the divide?: Capitalist reconstruction and the global agro-food system. The Journal of Peasant Studies, 22:1-49.

Hall, T. 1996. Legislative outlook. AgConsultant, 52:15.

Huffman, W.E., and R. Evenson. 1993. Science for Agriculture: A Long Term Perspective. Iowa State University Press.

MaCrae, R., J. Henning, and S. Hill. 1993. Strategies to overcome barriers to the development of sustainable agriculture in Canada: The role of agribusiness. Journal of Agricultural and Environmental Ethics, 21-49.

McConnell, G. 1969. The Decline of Agrarian Democracy. New York: Atheneum.

McMichael, P. (ed.) 1994. The Global Restructuring of Agro-Food Systems. Ithaca: Cornell University Press.

National Alliance of Independent Crop Consultants (NAICC). 1995. Statement by NAICC to the USDA Farm Bill Task Force Conservation Subgroup, August 10,1994.

National Research Council (NRC). 1995. Colleges of Agriculture at the Land Grant Universities: A Profile. Washington, D.C.: National Academy Press.

National Research Council (NRC). 1996. Colleges of Agriculture at the Land Grant Universities: Public Service and Public Policy. Washington, D.C.: National Academy Press.

Postelwaite A., D. Parker, and D. Zilberman. 1993. The advent of biotechnology and technology transfer in agriculture. Technological Forecasting and Social Change, 43:271-287.

Rivera, W.M., and D.J. Gustafson (eds.). 1991. Agricultural Extension: Worldwide Institutional Evolution and Forces for Change. Amsterdam: Elsevier.

Ruttan, V.W. 1982. Agricultural Research Policy. Minneapolis: University of Minnesota Press.

Schertz, L., and L. Daft (eds.) 1994. Food and Agricultural Markets: The Quiet Revolution. Report 270. Washington, DC: National Planning Association.

van der Ploeg, J.D. 1993. Rural sociology and the new agrarian question. Sociologia Ruralis 33(2):240-260.

Watts, M. 1994. Development II: The privatization of everything? 1994. Progress in Human Geography 18 (3):371-384.

Watts, S. 1996. CCA: How it all began. AgConsultant, 52:20.

Weaver, R.D. (ed.) 1993. U.S. Agricultural Research: Strategic Challenges and Options. Bethesda, MD: Agricultural Research Institute.

Wells, M. 1996. Strawberry Fields: Politics, Class, and Work in California Agriculture. Ithaca: Cornell University Press

Wolf, S. 1996. Privatization of crop production information service markets: Spatial variation and policy implications. Unpublished Ph.D. Thesis, University of Wisconsin-Madison, Institute for Environmental Studies.

Wolf, S. 1995. Cropping systems and conservation policy: The roles of agrichemical dealers and independent crop consultants. Journal of Soil and Water Conservation 50 (3):263-70.

Section One

STATUS AND TRENDS IN AGRICULTURAL INFORMATION

1

Institutional Relations in Agricultural Information: Transitions and Consequences

Steven A. Wolf

This chapter provides an overview of the questions addressed in this book. The first section of the paper identifies elements of the social, economic, and institutional environment in which information needs in agriculture are changing. The second section of the paper describes the economic utility of information and provides a caveat regarding overly generalized visions of agricultural development trajectories. The final section reviews a series of problematic issues which arise from changing patterns of public and private sector responsibilities.

SOCIAL AND ECONOMIC CONTEXT OF PRIVATIZATION OF AGRICULTURAL INFORMATION

Assessment of contemporary change in agricultural systems is supported through adoption of a historical and an interdisciplinary perspective. Here I briefly identify a set of six contextual factors that inform our understanding of information market development (i.e., growth in demand for and supply of information) and the shifting pattern of public-private institutional relations. The six factors are: the structure of farm and agribusiness sectors and industrial organization,

This introductory chapter draws upon contributions of other authors in this volume.

ISBN 1-57444-104-3/98/$0.00/$.50

technological innovation and human capital development, competitive markets, product differentiation, environmental and agroecological considerations, and decreased public investment in research and extension.

The dominance of agriculture production by fewer, larger businesses, and the global reach of these businesses, has dramatically changed the nature of agro-industry and state relations in agriculture. We now have a situation in which a small number of firms control elements of production and, more directly, the global marketing of a growing number of agricultural commodities (Heffernan and Constance 1994). The structure of agriculture and agribusiness sectors are, obviously, different from that of the days of the 1862 and 1890 Morrill Acts (establishment of the land grant university system), the 1887 Hatch Act (establishment of the State Agricultural Experiment Station system), and the 1914 Smith-Lever Act (establishment of Extension), major programs designed to promote scientific agriculture through the use of public resources. Modern structure is also largely different from that of the New Deal era, the period of origin of many contemporary policies, institutions, and the ideology upon which much of current state relations rest (Bergland 1981; Browne et al. 1992; Fite 1981; McConnell 1969). The structure of today is also quite different from that of twenty years ago, the period during which global restructuring of agro-food systems accelerated rapidly (Heffernan and Constance 1994; Little and Watts 1994; McMichael 1994; Schertz and Daft 1994). The emergence of highly capitalized, specialized farms, vertically integrated arrangements, and linkages between farms, input industries, and marketing institutions (i.e., vertical coordination) presents a situation in which the largest farms service their information needs internally (i.e., maintain in-house expertise) and a growing number of large and medium-size farms are relying on private sector consultants. Given changing structure and patterns of industrial organization, the traditional model of agricultural research and technology transfer, and Extension's role in that process, requires re-examination.

The flip side of concentration of production in agriculture and coordination of agro-food regimes is recognition of the social, economic, and environmental implications of large numbers of part-time and/or small farmers. Heterogeneity within the farm sector, for example, the differences in information needs and resources between highly capitalized, vertically integrated farming operations and small farmers, many of whom generate more income off-farm than on-farm, has led to a situation where monolithic theories, policies, programs, and institutions are encountering serious difficulty.

A second factor that must be integrated into a portrait of emergence of privatized information markets is that of technological innovation and enhancement of human capital. Development of new technologies, and increases in access combined with decreases in price of existing technologies, make it possible to develop "context-specific and decision-focused information" and communicate such information quickly and inexpensively (Boehlje in this volume). The ability to economically generate highly customized information products rather than generic sets of regional recommendations represents new levels of precision (e.g., Fortin and Pierce this volume). Computer-aided databases, monitoring equipment, and precision farming present opportunities for producers to capture information from their own operations rather than having to import and adapt information from off-site sources such as regional experiment stations. Innovations in communications, most notably the internet and cellular technologies, may transform the geography of information flow and further blur the distinction between rural and non-rural activity. Biotechnology, through production of new agricultural commodities (e.g., low fat oilseeds Stacey (1994)) and through changes in farming and processing of agricultural products, creates enormous information-related challenges and opportunities (Zilberman et al. in this volume).

Development of more sophisticated tools is linked to growing sophistication of workers. Farmers and farm service workers applying these technologies are more highly educated than in previous eras and now have the capacity and appetite to handle more, and more sophisticated, types of information. The expansion of information technology combined with development of human capital has created a new consumer audience for agricultural information, a new set of products, and new patterns of interaction between information suppliers and their clients (Sonka and Coaldrake 1996).

Third, while competitive markets in agriculture are clearly not new, market liberalization and deregulation trends as represented by the North American Free Trade Agreement (NAFTA), the General Agreement on Tariffs and Trade (GATT), and the 1996 Farm Bill (FAIR Act) indicate that production efficiency concerns are likely to increase in salience. As trade barriers relax, domestic subsidies are lowered, and larger numbers of producers compete for market share, low cost production will, more than ever, be a determining factor as to who wins and who loses. Additionally, the internationalization of agriculture—emerging geographic pattern of food and industrial manufacturing and consumption featuring corporate strategies of substitutionism (Goodman 1991), flexible production, and global sourcing of inputs and marketing of outputs (Bonanno 1994; McMichael 1994)—has exacerbated competitive pressures. In this globally competitive environment, farmers and agribusinesses have great incentive to capture the benefits of information and "shield" information from other firms (Holt in this volume; Postelwait et al. 1993).

Fourth, economic winners in agriculture are increasingly determined on the basis of their ability to create and participate in "differentiated" product markets (Schertz and Daft 1994). Product differentiation—for example, organic strawberries, high protein wheat, or free-range chicken—represents a means of adding value to a product through separating oneself from the large number of producers of fungible commodities. In essence, differentiation is an entrance into a market with fewer producers and therefore greater opportunity to set prices or benefit from higher prices. These value-added marketing strategies are popular at a time when basic commodity production is economically unattractive due to substantial inelasticity of demand for "raw" commodities and global sourcing of increasingly substitutable inputs. As retail consumers in many developed countries are willing and able to pay a premium for brand name products, and intermediate consumers (i.e., processors) have become increasingly sensitive to variation in feedstock quality (Greenwalt et al. 1995), production and marketing systems have developed around these identity preserved, post-Fordist products.[1] Interest in differentiated products has fueled demand and supply of specialized information as producers seek to develop, preserve, and enhance valued traits in their products.

The concept of differentiated products applies to farm inputs as well as output. Producers and input suppliers are increasingly focused on specific inputs rather than generic inputs. For example, growers are seeking, and suppliers are marketing, "super-fine" lime rather than generic lime (more surface area accelerates and enhances change in soil pH). Information and service markets are becoming similarly differentiated as more producers are paying specialized consultants as they see value in purchasing what they perceive to be higher quality information. Value-added processes are, of course, basic to marketing and are not new in agriculture. Such strategies are, however, receiving increased attention.

Fifth, responses to local and global environmental degradation (NRC 1989, 1992, 1993) and knowledge regarding the relationship between diet and health are substantially important for understanding the present course of agricultural development. In developed nations, food security no longer drives state intervention (Bunney in this volume). Given conditions of relative affluence, questions of ecology, human health, animal rights, and rural aesthetics have assumed great importance. Food safety, worker protection, and rural residents' health concerns pose liability challenges that shape demand for information. Increased knowledge of risks to humans and ecosystems, combined with sustained public support for environmentally-oriented intervention, influence the organization of production systems, the nature of emerging production technologies, and the research agendas of both public and private sector institutions. Demand for environmental accountability in agricultural production systems will likely be a singularly important justification for continued, and potentially enhanced, public sector investment in agriculture. In addition to information needs generated by resource conservation requirements, agroecological changes have created new demand for production information. Changes in soil quality, soil compaction, pest resistance, falling groundwater tables, salinization, and the potential for climate change represent emergent challenges that shape current and future information requirements in agriculture.

The sixth contextual factor addressed in this survey is that of declining public investment in agricultural research and extension. Despite firm documentation that the social benefits as measured by total financial returns to agricultural research exceed many if not most forms of public investment, it is often argued that the United States under-invests in such activity (ERS 1995a; Huffman and Evenson 1993; Ruttan 1982). Between 1960 and 1980, public (federal and state) funding exceeded private investment by a relatively narrow margin. Public funding for research was surpassed by private investments around 1980. After 1980, the amount by which private funding exceeded public funding grew sharply (Bonnen 1983; ERS 1995a). In 1992, public expenditures were roughly $2.5 billion and private expenditures were estimated, conservatively, at $3.7 billion (ERS 1995a).

In addition to relative declines in public sector engagement in research, public extension is playing a less active role in direct delivery of crop production information to farmers in the United States (Lambur 1989; NRC 1996; Wolf 1995) and in other developed nations (Rivera and Gustafson 1991). In the U.S. in 1992, total funds for Extension from federal, state, and local sources for agricultural and non-agricultural projects combined was roughly $1.4 billion (NRC 1995).[2] The most coherent explanation for a decline in Extension's role in providing production-oriented information is provided by Buttel (1991) who attributes these developments to the changing structure of the farm sector, the unraveling of the political constituency that has traditionally supported such public investment, and broadening of the mandate of Extension to include resource conservation in addition to yield-enhancing and cost-reducing information. The privatization of research has also contributed to the marginalization of Extension, as private sector technology transfer and education agents have direct access to the increasing volume of knowledge, information, and technologies produced outside of the land grant and USDA complex (Buttel 1991).

The general pattern of public fiscal austerity combined with competition for public resources among fragmented and largely urban/suburban interest groups, does not bode well for public support for investment in research and technology transfer. This attitude is exacerbated by public perceptions, right or wrong, that agriculture is industrialized—not exceptional relative to other economic sectors—and is, therefore, capable of funding and directing its own research and extension.

STRATEGIC APPLICATION OF INFORMATION IN AGRICULTURE

The six factors identified in the previous discussion constitute the macro context in which information has emerged as a critical and contested resource in agriculture (CARD 1995). At both the level of regionally, internationally, and globally linked agro-food commodity subsectors and the level of an individual firm (e.g., farm, agrichemical dealer, independent consultant, food processor, retailer), the value of information as a strategic competitive resource can be expressed through consideration of uses for information. Five interrelated uses are identified here.

1. Information supports effective resource allocation decisions—land, capital, labor, management.
2. Information serves to coordinate activity within and between nodes in complex commodity chains. Information and communication supports just-in-time/flexible production within multi-institutional, geographically disaggregated production processes.
3. Information supports development, enhancement, and preservation of product and production process attributes desired in the marketplace (e.g., size, moisture content, organically produced). Production monitoring and intensive management make it possible to add value to products through identity preservation and insure the integrity of differentiated products.
4. Information supports risk management planning and accountability. Firms seek to manage liability by gathering high quality information, maintaining records, and outsourcing specific aspects of production processes. Information management allows firms to negotiate effectively with allied firms, insurers, lenders, buyers, and regulators through use of comprehensive, verifiable (transparent) and easily accessible records.
5. Information supports continual improvement over time. Continuous monitoring of operations allows firms to seek productivity enhancing opportunities and engage in experimentation as part of production processes. Analysis of production performance data represents in-house research and allows firms to internalize (i.e., avoid sharing) information and resulting production innovations.

Information, as represented by these five applications, provides a means through which firms and production sub-sectors are able to react to and capitalize on macro-level developments described in the previous section. The strategic value of information is derived, in part, from restructuring and reorganization, technological change, intense market competition associated with globalization, fragmentation of consumer markets, environmental risks, and privatization of research and extension. Institutional innovations, such as vertical coordination and integration, and technological innovations, such as precision farming, internet, and biotechnology, are examples of strategies and tools that utilize information to meet these contemporary production and marketing challenges (Streeter et al. 1991).[3]

Movement towards internalized sourcing of information and proprietary information is consistent with patterns of vertical coordination and larger patterns of restructuring (Bonnano et al. 1992; Goodman and Watts 1994; McMichael 1992). Vertical coordination, commonly pursued through production contracting, is now a major trend in agro-food systems and a growing empirical focus in agrarian studies (e.g., Little and Watts 1994). As Watts states (1996:233), "(g)iven the concerns with quality and market niches, contract production is a fundamental way in which the division of labour of ... global commodity systems is organized." While Watts is describing down-stream coordination (i.e., producers' marketing arrangements) and I am addressing up-stream coordination (i.e., producers' sourcing of information), the arguments are essentially parallel. Given interest in the quality of farm products and development of specialized markets, flexibility, and competitive advantage, a system for development and application of specialized production information has developed that features more internal sourcing, and more intensive coordination between on-farm production and off-farm interests. Identifying the connection between privatization of information and coordination of production (e.g., growth in contract production) is central to recognition of how this book contributes to the agro-food restructuring literature (Goodman and Watts 1994; McMichael 1994).

Additionally, this book contributes to the restructuring literature by focusing on changing patterns of production and application of farm-level production inputs. In my view, the restructuring literature has overwhelmingly been focused on down-stream, post-harvest aspects of the process of integration of farms into industrial agro-food systems (e.g., Schertz and Daft 1994). This focus limits understanding of how restructuring specifically impacts production practices (Zilberman et al. 1994), on-farm labor processes (Friedland et al. 1981; Wells 1996), and environmental quality (Wolf and Nowak 1995). This book attempts a more balanced treatment. Emphasis in this collection of papers is on linkages between farm-level production and pre-harvest segments of commodity chains. Given interest in the environmental and ecological impacts of agriculture and food systems, analysis of the impact of restructuring on farming practices and production decision-making processes has obvious utility.

The treatment presented here of emerging patterns of development and application of information emphasizes coherence, perhaps a false coherence, between macrostructural trends and localized behavior of firms and farm-level producers (Buttel 1994; Goodman and Watts 1994; van der Ploeg 1993; Wells 1996). Conversations with most farmers and many agribusiness employees about their relationships with elements of the global food system would yield very different results concerning connectivity between segments of commodity chains and levels of organization. The size, diversity, complexity, and local emeddedness of agriculture poses serious theoretical and empirical problems for analysts. In order to develop a richer understanding of public and private institutional relations, micro-level studies of production, distribution, and application of information will be required.

Figure 1 represents a comical portrait of the juxtaposition of post-industrial concepts such as product life cycle analysis, vertical coordination, biotechnology, and precision farming with simple commodity production firmly situated in a rural context. This cartoon illustrates the contradictions and discontinuities that emerge from analysis of farming systems as a component of global agro-food regimes.

IMPLICATIONS OF PRIVATIZATION OF INFORMATION

Here I outline a series of risks and opportunities associated with privatization of development and delivery of information in agriculture. My objective is to briefly survey a series of considerations. Comments are directed toward environmental impacts, conflict of interest, impacts on innovation, economic and social impacts, labor process and the social relations of production, market failures, and finally, a set of opportunities presented by restructuring of institutional relations regulating agricultural information.

Environmental impacts

As many authors in this volume indicate, at the level of farming systems and communities—the agricultural and ecosystem interface—privatization of agricultural information translates into less reliance on Extension and more reliance on private technical consultants, or, in the case of very large farms, in-house expertise. The integration of private consultants into field- and farm-level decision making positions these firms to affect the environmental performance of crop-

Figure 1. Farming in a globalizing agro-food system. S. Wolf. Artwork by Brian Obach.

ping systems, particularly with respect to fertilizer, pesticide, and water use efficiency (Hoffman 1993; Wolf 1995; Zilberman et al. 1994). These firms contribute to, constrain, and in some production systems define the sophistication of farmer's management of fertilizers and pesticides. Consultants provide input use recommendations based on analytic and diagnostic services of varying quality and type. The rigor and sophistication with which data are collected, synthesized, and applied in conjunction with the scale at which these processes occur determine fertilizer and pesticide use efficiencies. Enhanced application of site-specific information represents a tremendously important opportunity to mitigate resource degradation associated with agriculture. A great deal of the agroecology literature is focused on enhancing the site-specificity of monitoring and the development of an information-intensive agriculture. In this specific context, greater use of technical consultants in farming systems is consistent with an agroecological perspective.[4]

The success of efforts to assist producers in implementing irrigation scheduling, integrated pest management (IPM), realistic nutrient budgets, precision farming, and other pollution-preventing technologies that conserve resources while boosting productivity will be determined, in large part, by private consultants.[5] The increasing importance of commercial infrastructure in shaping cropping practices raises a series of concerns regarding the type and quality (i.e., accuracy, timeliness, price) of information available to producers and the spatial heterogeneity likely to characterize information markets.

Conflict of interest

The quality of information available to producers, specifically impartiality of fertilizer and pesticide recommendations made by private sector consultants, will be a critical determinant of the efficiency of farm chemical use and the environmental performance of production systems. To the extent that consultants stand to receive financial or other benefits (e.g., gifts, community status) by making recommendations that lead to over-application or improper selection of products, a conflict of interest exists. While we do not know the extent to which biased information impacts production activity, the issue is consistently raised by a variety of interest groups. Growing roles for private consultants are occurring at the same time that county-level agricultural Extension resources are shrinking. Reduced community-level presence of Extension makes it less convenient and affordable for farmers to confirm the accuracy of recommendations through a second opinion.

Environmentalists have condemned agrichemical dealers for playing more active roles in determining farmer's fertilizer and pesticide management practices on grounds that it is a case of leaving the fox to guard the hen-house (Dysart

1994). While there are grounds for critical appraisal of information provided by self-interested individuals and firms, it is likely that the problem is more complex than one of dealers versus independent consultants. Independent consultants provide information for a fee and sell no products of any sort. As discussed by Wolf in this volume, the label "independent", as defined simply by a consultant not employed directly by a farm chemical dealer, will be an inadequate measure of the extent to which an individual consultant may be subject to a conflict of interest. Efforts to build and maintain professionalism in the consulting industry by the USDA, National Alliance of Crop Consultants, the Certified Crop Advisor program, and other institutions are of critical importance (e.g., AgConsultant 1996:14).

There is an additional form of bias operating at a macro institutional level. Findings of a body of research reviewed by Buttel et al. (1990:135-144) support several critical statements. The organization of agronomic knowledge production within the present structure supports a research agenda and educational system biased toward generation of knowledge and technology compatible with industrial modes of production, meaning heavy dependence on purchased inputs including farm chemicals. Individual researchers, the universities and agencies in which they work, and the larger community of public agricultural research institutions devote substantial resources to retaining the political and financial support of industry. In reviewing the pattern of allocation of research funds (NRC 1995:118), it is appropriate to scrutinize the extent to which the current structure serves to reorient agriculture toward a more ecologically-intensive mode of production or toward reproduction of existing arrangements.

Implications for innovation

In considering privatization of information within the context of the related trend of privatization of agricultural research (ERS 1995a), it is important to question whether these trends jeopardize the capacity of agriculture to explore alternatives and depart from a system of production largely dependent on fossil fuels and industrially produced inputs (e.g., genetics, farm chemicals). Privatization potentially compromises the long-term innovative capacity of agriculture as private firms engaged in research are expected to act in ways that increase return on capital investment and insure sustained demand for their products and services. This relationship suggests that farming systems and component technologies featuring lowering or elimination of demand for off-farm inputs are not likely to be high on the research and extension agenda of for-profit firms engaged in farm input sales (See Keeney and Vorley in this volume).

Privatization of agricultural research raises questions about who will conduct basic and speculative (high risk) research, who will develop products for small markets, and who will conduct social science and policy-relevant research, farm management (integrative, systems oriented) research, and agroecological research. Privatization is one constraint or a larger set of inter-related factors contributing to under-production of specific types of research products and Extension functions. Structural issues associated with land grant university research environments, specifically reward structures for researchers (Bird 1996) and transaction costs of doing interdisciplinary, problem-focused work, limit applied knowledge production and the nature of university outreach activities (Kalaitzandonakes and Bullock in this volume; NRC 1989:14). As universities struggle to redefine patterns of engagement and develop roles that are complementary to those of the private sector, addressing constraints to interdisciplinary teaching, research, and extension will be of great importance.

Economic and social impacts

Commercialization of information markets is likely to contribute to structural change in agriculture and distributional inequities (NRC 1996:95). As information and information management emerge as growing, variable costs, agricultural production becomes more capital intense. Capital intensification creates barriers to entry and concentration of assets among existing, larger firms.

Farm structure will be impacted by community-level, social interactions as well. Consultants are expected to use sliding fee scales favoring large farms to attract and retain large farm clients in order to minimize their transaction costs. Additionally, leading consultants are expected to ally themselves strategically with that segment of the farm sector likely to be around in the decades that lie ahead. As a source of strategic competitive advantage, firms able to access and pay for better information—the "information elite"— are expected to lower their production costs and negotiate more effectively with vendors and consumers, accelerating the ongoing structural transformation.[6]

In addition to concentrating assets among fewer farms, privatization implies that producers in intensive production regions will enjoy differential access to "world-class information", leading to geographical concentration of production in areas in which the information infrastructure is well developed. Areas with less sophisticated information development and delivery capabilities, and less well-trained consultants, are expected to find themselves at a competitive disadvantage due to emerging information-based competition. As the information infrastructure grows in importance, clustering of production and processing facilities is likely. Such clustering presents environmental and ecological risks and creates community economic dependence on single industries subject to periodic booms and busts (Purvis in this volume).

Labor process and the social relations of production

Appropriation of information development functions and integration of agribusiness into farming through privatization of extension is consistent with a pattern of industrialization of agriculture (Goodman et al. 1987). More intense involvement of off-farm interests in more types of on-farm activities indicates continued transformation of the social relations of agricultural production. The appropriation and commodification of information[7]—assignation of property rights to goods previously allocated through a system of "open" access—is expected to have significant implications for the social relations of production (Goe 1986). By establishing logistical and legal control (i.e., property rights) over a fundamental component of production, in this case information, individuals and firms providing information establish imbalanced (i.e., asymmetrical) dependency relations with producers (Martinson and Campbell 1980).

Integration of off-farm consultants into farm and field level decision-making potentially leads to subordination of farmers by agribusiness and homogenization of the agricultural labor process. The issue of homogenization of labor process of farmers is of particular interest when considering the role of indigenous or ethno-scientific knowledge in agriculture (McCallister 1996) and increasing prevalence of computers in society. In an age of subcontracting or outsourcing of component tasks that comprise production, farmers become managers and coordinators of others' labor. Farmers are not necessarily subject to loss of skill. More accurately, their traditional skills are replaced with those patterned after industrial modes of production—negotiation, human relations, and finance.

Privatization of information is a product of and contributor to the industrialization of agriculture. Increasing the intensity of information use in cropping systems through development of commercial consulting and application of sophisticated information technologies is consistent with the long run pattern of agriculture losing its distinctive qualities with respect to manufacturing and other industrial sectors.

Market failures

Privatization of information transfer implies that the content of the information, terms of transactions between producers and consumers of information, and resultant resource management behaviors arising out of the information are expected to be less responsive to policy signals and to be increasingly sensitive to market signals. In this context, market failures are of great concern. I am using the term market failure to refer to the impact of property rights regimes on production of specific goods and inducement of specific behaviors. In addition to those mentioned previously, classes of market failure of particular concern include insufficient commitment and capability to provide information on natural resource conservation and disembodied, agroecological production techniques (e.g., cultural practices such as crop rotation and cover cropping), differential access to information among segments of the farm sector, and constraints on research and extension. Questions surrounding who will perform those service functions deemed unprofitable or strategically unwise by private industry, and with what resources, will become increasingly important in an era of privatization of information (Boehlje 1994).

In addition to market failures, institutional gaps impacting overall system performance potentially arise as privatization proceeds. Three potential gaps are discussed briefly: connectivity between research and the field, public access to data, and private sector accountability.

Severing the feedback function in the land grant university system whereby outreach (extension) and research inform one another, may lead to inefficiencies and inequities (Bonnen 1983; Buttel 1991). In addition to concerns surrounding efficient technology transfer and the relevance of research products, decreased field level presence associated with diminished Extension service may lower the visibility of public institutions. Generating recognition of the contributions flowing from public research and extension, maintaining a politically active constituency for continued public sector engagement, and ultimately, preserving local, state, and federal fiscal appropriations is a critical challenge facing public institutions. "Getting credit" becomes increasingly difficult as public institutions move into background roles that support private sector development and delivery of information rather than direct face-to-face extension.

The second and third institutional gaps—lack of public sector access to information and insuring private sector accountability—involve public agencies collecting and controlling fewer data, and data of lower quality than that of private firms. Given that private firms have few incentives to share data within the current system and public agency capabilities with respect to data collection are in question, there is reason to expect that public institutions and private, policy-oriented researchers may have problems accessing state-of-the-art information (AAEA Data Task Force, 1997). When data are not publicly accessible, ability to conduct resource inventories, detect market failures, and undertake policy analysis will be curtailed. Further, the system of checks and balances that characterize relations between the government, industry, and not-for-profit watchdog, advocacy, and educational groups will be compromised. Accountability of individuals and corporations will be increasingly difficult to enforce as information is internalized by individuals and firms. In addition to a diminished capacity to police behavior, diagnose problems, and develop mitigation strategies, these gaps suggest that the public sector's capability to intervene will be limited through a process of privatization.

Opportunity through privatization

While market failures and institutional gaps tend to attract significant attention, privatization of information presents opportunities for enhanced investment, efficiency gains (narrowly defined), and accelerated innovation. For example, development of strengthened property rights on products of biotechnology has led to accelerated investment by private firms (ERS 1995b). Enhanced market development of analytic and diagnostic services such as soil testing, integrated pest management, animal health, precision technologies for fertilizer, pesticide, and water inputs will likely speed farm-level adoption. Competition among suppliers engaged in open markets will serve to bolster the quality of informational services available to producers.

Refocusing public institutions such as Extension, the Natural Resources Conservation Service, the Agricultural Research Service, and State Agricultural Experiment Station system and other public agencies away from activities private sector firms are willing to undertake may support reallocation of funding toward traditionally under-funded projects. Targets of enhanced public investment could include basic research, natural resource conservation, ecological restoration, and development and implementation of a national rural policy (Browne et al. 1992; NRC 1996).[8] To the extent there is redundancy between public and private investments, and the redundancy does not create valued benefits, elimination or reduction of public programs is consistent with common sense.

The greatest opportunities evident from an analysis of current trends in institutional relations governing agricultural information lie in public-private partnership. In the introduction of this book I identified Extension, the Certified Crop Advisor program, and university-based offices of technology transfer as examples of public-private collaboration. The remaining chapters of this book address a large number of other examples and provide a conceptual framework within which to pursue institutional innovation. However, it should be noted that the remaining chapters also identify a series of sobering constraints on the applicability of public-private collaboration as currently conceptualized.

NOTES

[1] Buttel (1992:4) has characterized emergence of an agriculture focused, at least in part, on post-Fordist commodities (See Goodman and Watts (1994) for a critical discussion of this terminology) as consistent with a larger pattern of income stratification in the economies of developed nations. Unlike previous eras in which industrial production was targeted towards mass markets (i.e., Fordism), we now observe targeting of upper-income consumers, those consumers who have seen their incomes rise in recent decades, as represented by proliferation of value-added, luxury products.

[2] I am not aware of any estimate of private agricultural extension investments. Such an analysis would have to confront the problem of discerning between research, extension, and marketing expenditures.

[3] A growing segment of agriculture is utilizing a different set of strategies and tools for coping with changes in production and marketing environments. Alternative farming systems featuring low-input production, specialty crops, direct sales to consumers, on-farm processing, and other innovative strategies represent an important element of heterogeneity in agriculture (e.g., The Center for Rural Affairs' on-farm value added initiative). These out-of-the-mainstream strategies are poorly addressed in this volume yet constitute important topics of research.

[4] Consistency between privatization of information and agroecological conceptions of appropriate institutional reforms quickly breaks down. See Keeney and Vorley (in this volume) for a discussion of some aspects of disjuncture.

[5] My comments focus on the roles of off-farm consultants and therefore tends to underreport production systems predicated on more independent or cooperative information development strategies. Neighbors and relatives are important sources of information in most production systems. Also, farm magazines, other media, public sector institutions such as Extension, and farmer-to-farmer networks are recognized as important sources of information.

[6] Here we focus on production information, but differential access to marketing information creates a similar opportunity for structural change. Public sector provision of commodity price data in the U.S. has been identified as a key factor in limiting concentration in agriculture and maintaining a competitive structure at the farm-level (Zilberman et al. in this volume).

[7] Some would argue that a key process described in this book, movement towards customized, differentiated information and away from interchangeable, unfocused information, constitutes de-commodification of information, quite the opposite of commodification. This confusion of terms is the result of cross-disciplinary communication problems. I am using the term commodification in a social context to refer to a process of appropriation (Goodman et al. 1987). In an economic context, it could be argued that de-commodification, the process of differentiating a product from a previously homogenous mass, is essential for establishment of property rights.

[8] Disengagement has potential benefits as the private sector may be positioned to address applied and adaptive research, freeing up the public sector research institutions to do more long-term, basic research. Boehlje (1994) has argued that such a division of responsibilities is logical according to the comparative advantage of the institutions involved.

REFERENCES

AAEA (American Agricultural Economics Association) Data Task Force. 1997. Report to USDA Economic Research Service and National Agricultural Statistics Service.

AgConsultant. 1996. Vol. 52, No. 4:14.

Bennett, C.F. 1994. Rationale for public funding of sustainable agriculture Extension programs. Paper presented at American Evaluation Association Annual Meeting. Vancouver, BC, Canada. November 1-5, 1995.

Bergland, R. 1981. A time to choose: Summary report on the structure of agriculture. Washington, DC: USDA.

Bird, E. 1996. Rewards for sustainable agriculture research and education. Consortium News No. 11. Consortium for Sustainable Agriculture Research and Education. Madison, WI.

Boehlje, M. 1994. Information: What is the public role? Purdue University Staff Paper 94-17.

Bonanno, A., L. Busch, W. Friedland, L. Gouveia, and E. Mingione, (eds.) 1994. From Columbus to Conagra: The Globalization of Agriculture and Food. Lawrence: University Press of Kansas.

Bonnen, J. 1983. Historical sources of U.S. agricultural productivity: Implications for R&D policy and social science research. American Journal of Agricultural Economics, Dec. 1983: 959-966.

Browne, W. P. , J. Skees, L. Swanson, P. Thompson, and L. Unneveher. 1992. Sacred Cows and Hot Potatoes: Agrarian Myths in Agricultural Policy. Boulder:Westview Press.

Buttel, F.H. 1994. Agricultural change, rural society and the state in the late twentieth century: Some theoretical observations. In Agricultural Restructuring and Rural Change, Europe, D. Symes and A.J. Jansen (eds.). The Netherlands: Wageningen Agricultural University.

Buttel, F.H. 1991. The restructuring of the American public agricultural research and technology transfer system: Implications for agricultural Extension. In Agricultural Extension: Worldwide Institutional Evolution and Forces for Change, W. Rivera and D. Gustafson (eds.). Amsterdam: Elsevier.

Buttel, F.H. 1992. Environmentalization: Origins, processes, and implications for rural social change. Rural Sociology 57:1-27.

Buttel, Frederick H., O. Larson, G. Gillespie. 1990. The Sociology of Agriculture. New York: Greenwood Press.

CARD (Center for Agriculture and Rural Development). 1995. National forum for agriculture: The power and politics of information. Des Moines, IA. February 27-28.

Cary, J. 1993. Changing foundations for government support of agricultural Extension in economically developed countries. Rural Sociology XXXIII(3/4), pp. 336-347.

Cochrane, W. 1979. The Development of American Agriculture. Minneapolis, MN: University of Minnesota Press.

Dysert, J. 1994. Green movement pushes to make crop advice illegal. Custom Applicator, Oct.

ERS (USDA/Economic Research Service). 1995a. Agricultural research and development: Public and private investments under alternative markets and institutions. Staff Paper 9517. Washington, DC: USDA.

ERS (USDA/Economic Research Service). 1995b. New crop varieties. Agricultural Resources and Environmental Indicators. RTD Update, 14.

Fite, G.C. 1981. American Farmers : The New Minority. Bloomington: Indiana University Press.

Ford, S.A., and E.M. Babb. 1989. Farmer sources and uses of information. Agribusiness, 5(5): 645-476.

Friedland, W., L. Busch, F. Buttel, and A. Rudy. 1991. Towards a New Political Economy of Agriculture. Boulder: Westview Press.

Friedland, W., A. Barton, and R. Thomas. 1981. Manufacturing Green Gold. New York: Cambridge University Press.

Goe, R.W. 1986. U.S. agriculture in an information society: Rural sociological research. Rural Sociology 6(2): 96-101.

Goodman, D. 1991. Some recent tendencies in the industrial reorganization of the agri-food system. In Towards a New Political Economy of Agriculture, W. Friedland, L. Busch, F. Buttel, and A. Rudy (eds.) Boulder: Westview Press.

Goodman, D., B. Sorj, and J. Wilkinson. 1987. From Farming to Biotechnology. Oxford: Basil Blackwell.

Goodman, D., and M. Watts. 1994. Reconfiguring the rural or fording the divide?: Capitalist reconstruction and the global agro-food system. The Journal of Peasant Studies, Vol. 22, No. 1:1-49.

Greenwalt, B., D. Schweikhardt, and M. Fairley. 1995. Quality and coordination in the food system: The case of Riceland Foods. In Teaching and Learning with Cases, S. Swinton (ed.). East Lansing: Michigan State University.

Heffernan, W. and D. Constance. 1994. Transantional corporations and the globalization of the food system. In From Columbus to Conagra: The Globalization of Agriculture and Food, Bonanno, A., L. Busch, W. Friedland, L. Gouveia, and E. Mingione (eds.). Lawrence: University Press of Kansas.

Hoffman W.L. 1993. Stemming the flow: Agrichemical dealers and pollution prevention. Environmental Working Group, Washington, DC.

Huffman, W.E., and R. Evenson. 1993. Science for Agriculture: A Long Term Perspective. Iowa State University Press.

Lambur, M., R. Kazmierczak, Jr., and E. Rajotte. 1989. Analysis of private consulting firms in integrated pest management. Bulletin of the Entomological Society of America, Spring 1989: 5-11.

Little, P. and M. Watts (eds.). 1994. Living Under Contract. Madison, WI. The University of Wisconsin Press

Martinson, O.B., and G.R. Campbell. 1980. Betwixt and between: Farmers and the marketing of agricultural inputs and outputs. In The Rural Sociology of Advanced Societies, F. Buttel and H. Newby (eds.). Montclair, NJ: Allanheld, Osmun & Co.

McCallister, Robert. 1996. How Wisconsin farmers understand and manage their soil landscapes: A site-specific people and place methodological analysis. Ph.D. Dissertation, University of Wisconsin-Madison.

McConnell, G. 1969. The Decline of Agrarian Democracy. New York: Atheneum

McMichael, P. (ed.) 1994. The Global Restructuring of Agro-Food Systems. Ithaca: Cornell University Press.

National Research Council (NRC). 1996. Colleges of Agriculture at the Land Grant Universities: Public service and public policy. Washington, DC: National Academy Press.

NRC (National Research Council). 1995. Colleges of Agriculture at the Land Grant Universities. Washington, DC: National Academy Press.

NRC (National Research Council). 1993. Soil and Water Quality: An Agenda for Agriculture. Board on Agriculture, Washington, DC: National Academy Press.

NRC (National Research Council). 1992. Global Environmental Change. Washington, DC: National Academy Press.

NRC (National Research Council). 1989. Alternative Agriculture. Washington, DC: National Academy Press.

NRCS (USDA Natural Resources Conservation Service). 1996. Data rich, knowledge poor. Report of a Blue Ribbon Panel to Chief Paul Johnson. Washington DC: Office of the NRCS Chief.

Postelwaite A., D. Parker, and D. Zilberman. 1993. The advent of biotechnology and technology transfer in agriculture. Technological Forecasting and Social Change, 43:271-287

Rivera, W.M., and D.J. Gustafson (eds.) 1991. Agricultural Extension: Worldwide Institutional Evolution and Forces for Change. Amsterdam: Elsevier.

Rivera, W.M., and J.W. Cary. (in press) "Privatising" agricultural Extension: Institutional changes in funding and delivery of agricultural extension. In Improving Agricultural Extension: A Reference Manual, B.E. Swanson (ed.). Rome: Food and agriculture organization.

Ruttan, V.W. 1982. Agricultural Research Policy. Minneapolis: University of Minnesota Press.

Schnitkey, G., M. Batte, E. Jones, and J. Botomogno. 1992. Information preferences of Ohio commercial farmers: Implications for Extension. American Journal of Agricultural Economics, May: 486-496.

Schertz, L., and L. Daft (eds.) 1994. Food and Agricultural Markets: The Quiet Revolution. Washington, DC: National Planning Association.

Sonka, S., and K. Coaldrake. 1996. CyberFarm: What does it look like? What does it mean. American Journal of Agricultural Economics, Vol. 78, No.4.

Stacey, R. 1994. Biotechnology and the globalization of the U.S. food industry. In Food and Agricultural Markets: The Quiet Revolution, L. Schertz, and L. Daft (eds.). Washington, DC: National Planning Association.

Streeter, D., S. Sonka, and M. Hudson. 1991. Information technology, coordination and competitiveness in the food and agribusiness sector. American Journal of Agricultural Economics 73(5).

van der Ploeg, J.D. 1993. Rural sociology and the new agrarian question. Sociologia Ruralis 33(2):240-260.

Watts, M. 1994. Development II: The privatization of everything? Progress in Human Geography 18 (3):371-384.

Wells, M. 1996. Strawberry Fields: Politics, Class, and Work in California Agriculture. Ithaca: Cornell University Press

Wolf, S. 1995. Cropping systems and conservation policy: The roles of agrichemical dealers and independent crop consultants. Journal of Soil and Water Conservation 50(3):263-70.

Wolf, S., and P. Nowak. 1995. Development of information intensive agrichemical management services in Wisconsin. Environmental Management Vol. 19 (3):371-382.

Zilberman, D., D. Sunding, M. Dobler, M. Campbell, and A. Manale. 1994. Who makes pesticide use decisions: Implications for policymakers. In Pesticide Use and Produce Quality. Oak Brook, IL: Farm Foundation.

2

Information and Technology Transfer in Agriculture: The Role of the Public and Private Sectors

Michael Boehlje

The system and mechanism by which farmers obtain new technology and information is changing dramatically. Private sector providers of information are expanding relative to the public sector, and information has become a source of strategic competitive advantage for both farm input and service providers. Information is becoming more detailed with the potential for increased accuracy and resolution. Dissemination technology has reduced the cost of accessing information, and will make real-time personalized messages available anytime, anywhere, anyplace. Information is becoming an increasingly important driver of control and structural change in the agricultural industry, and access to information and intellectual property rights are becoming increasing sources of conflict and controversy as information increases in value and that value can be captured by private sector firms. The policy issues with respect to the future role of the public sector, public/private sector linkages, distributional issues, access to information, and intellectual property rights in a global context are critical to the competitiveness of individual firms and regions and the rate of economic growth and development in agriculture and rural communities around the world.

The critical role of information and new technology as a source of competitive advantage and continuous improvement in business and financial performance has been recognized by farmers for years. U.S. farmers have been acclaimed worldwide as almost insatiable consumers of information and adopters of new technology. A significant factor in the dramatic productivity increase of U.S. ag-

ISBN 1-57444-104-3/98/$0.00/$.50

riculture during the past half century has been the information on new ideas and techniques that farmers have gleaned from both the public and private sector.

Knowledge and information has always been an important resource to business managers, but its relative importance has increased in recent years (Drucker 1992; Peters 1992). Whereas the physical resources of land, labor, and capital combined with a bit of knowledge and information were the key determinants of financial success in the past, the role of knowledge and information has and will likely become more important in the future for successful management of a farm business. Superior knowledge and information will be the cornerstone for success—it will enable the producer to obtain the physical resources of land, labor, and capital and combine them in an efficient manner. Knowledge and information about a broader and more complex set of issues—for example environmental and ecosystem dimensions of farming as well as production technology and institutional structures—is increasingly important for profitable and socially responsible farm operations.

This discussion will assess the impact of changes in where producers obtain information—more from private sector firms and less from public institutions. We will first discuss some critical concepts necessary to understand this change which we will refer to as the privatization of information. Then we will identify some key consequences of privatization. Finally, some of the critical policy issues of the privatization of information will be discussed. The discussion will focus most explicitly on information rather than technology transfer, although the majority of the arguments are applicable to both.

CRITICAL CONCEPTS

What is information?

The concept of information means different things to different people. In this discussion, we will distinguish between three important concepts—1) knowledge, 2) data, and 3) information. This distinction will not only assist in understanding the discussion that follows, but it can also be used to think about the role of the public sector.

Knowledge is the broad-based concepts, theories, principles and models that are necessary to understand a particular phenomena. Knowledge can be applied broadly across many sets of facts and circumstances or contexts. It is not data specific, but helps one sort through the vast quantities of data available to determine what is relevant. Knowledge assists one in determining what data is useful in analyzing and understanding a particular issue or making a specific decision.

Data are more specific than knowledge; they may be individual numbers or observations or data might be an individual idea or concept. Data can be quantitative or qualitative in nature. At the extremes data are distinguishable from knowledge in that data are specific while knowledge is general. Clearly this clean distinction becomes fuzzy at times.

Information is different from data or knowledge in two important dimensions: first it is context specific, and second it is decision focused. In essence, if knowledge and data are combined and applied to a specific context (for example, a specific crop and parcel of land) and a specific decision (the proper level of fertilizer to apply to obtain a particular yield of a particular crop), they are transformed into information. Information becomes more valuable as it results in improved decision making and better physical and financial performance.

Information has many attributes. It must be timely—appropriate to the decision context and not out-of-date. It must be technically accurate and scientifically sound. It must be objective and unbiased, and/or value judgments must be explicitly identified. It must be complete (as opposed to partial) so as to be useful in a decision, or its partial or incomplete nature must be clearly specified. It must be understandable—communicated in such a way that the user can comprehend it. Finally it must be convenient—available when and where and at what time the user needs or wants it. These attributes will determine the value of information.

Although these definitions may differ somewhat from those provided by other authors, they will provide a useful context for the discussion of the privatization of information and technology transfer in the increasingly complex information driven agricultural industry.

Incentives to privatize

The increased context specificity and decision focused nature of information in recent years has increased its value. As information becomes more valuable, the incentive for the private sector to provide that information and capture some of that value increases. Consequently, growth in private sector data gathering and information service firms is not surprising, given the growing value of information.

Information can also be a significant source of strategic competitive advantage. Those firms that can obtain superior information can act on that information and improve their performance compared to those firms with inadequate access to the latest and best information and technology. Thus superior (better in terms of context specificity and decision focus) information is a source of competitive advantage for the supplier of that information—allowing him/her to extract value or income from the user of that information by charging fees or maintaining or improving related product sales. It also provides a competitive advantage for the user of that information—in this case the producer—in terms of better performance and higher profits compared to other producers.

Not only is the relative role and importance of information and knowledge changing, the sources of that information for farmers is also going through a transformation process. Farmers have access to more information from the private sector (or from internal sources on the part of large scale or integrated producers) and less from the public sector. In many cases, providers of key farm inputs such as pharmaceuticals and chemicals have become critical suppliers of information along with those inputs, leaving the traditional Extension Service and land grant/USDA complex at a significant disadvantage in terms of providing the latest technology and information. In some geographic regions, larger and more educated producers are becoming a larger proportion of U.S. producers, and rate traditional public sector information sources such as county extension agents and even university specialists significantly lower than many other sources of information for production, marketing, or financial decisions (Ortmann et al. 1993). These dramatic changes—both in the importance of information and the preferred provider of that information—raise a number of questions about the changing role of the public sector in the knowledge, data, and information industries.

Intellectual property rights

A critical concept that will be useful in understanding who gains and who loses with the privatization of information is that of property rights—both rights in intellectual property and rights to the data. Because of the extensive if not dominant role of the public sector in providing information to producers in the past, information was common property and individual data, if shared, was generally not proprietary and confidential. But as an increasing amount of the data in production agriculture is gathered by suppliers, consultants, and service firms, and the private sector plays a larger role in providing both data and information, private property rights replace common property concepts. Private property rights enable individuals who have those rights to capture value—to extract rents from those who use that property. Consequently, with the growing privatization of the information markets, intense debates and litigation will occur as to who has the intellectual property rights in systems of monitoring and interpreting data, and who has property rights to the data and thus can control its accessibility. Those who have property rights in information and data will benefit more from that information than those who do not have those property rights. Furthermore, differential values based on the exclusivity or other dimensions of property rights may impact who receives the most valuable information. For example, in smaller farm operations the value of data may be much lower than the cost of collecting that data, and thus there is no incentive for profit motivated firms to collect and analyze that data. Thus, smaller firms would be at an disadvantage as the privatization of information process increasingly emphasizes larger firms where relatively more rent can be captured from the property rights in data and information.

Technological innovation vs. institutional innovation

Innovation in agriculture is fundamentally of two forms—technological innovation and institutional innovation. Technological innovation is familiar to producers; it is the new genetics, the new nutrition and feed regimes, the new tillage techniques, the new types of livestock production facilities, etc. Institutional innovation is less familiar and more threatening. In essence, institutional innovation involves new ways of doing business: new business linkages such as contract production and vertical integration; new financing and asset control strategies such as leasing and borrowing from the machinery dealer rather than a bank or traditional lender; new forms of businesses such as limited liability companies and closed cooperatives; new organizational structures that include more hired employees and fewer family members, etc. Institutional innovation may be more threatening because it often involves changes in relationships as well as new ideas that may challenge fundamental values such as independence. Much of the structural change occurring currently in Midwestern agriculture involves significant institutional innovations in combination with technological innovation; the increased vertical linkages and industrialization of the pork sector is but one example.

The type of information required to assess and select alternative institutional innovations is somewhat different from that required for technological innovations. From a firm level perspective, technical innovations can generally be adequately assessed with accurate information on physical performance characteristics and financial and economic costs and benefits. Because institutional innovation changes relationships and challenges values, additional information on life-style consequences is necessary. These life-style consequences are often dependent on the personality, value systems, and personal and social circumstances of the decision-maker. For example, the decision of whether or not to produce grain under contract or as an independent producer is not driven by cost and revenue alone, but attitudes towards risk and independence in decision making as well. Consequently, not only is more information required by the decision-maker to make choices on alternative institutional innovations, the type of information is also different. The role of values and personality of the decision-maker in the calculation mean that recommendations and generic prescriptions are less useful.

Instead, information on how to analyze the alternatives; how to combine physical and financial information with values, personality characteristics and social circumstances; how to resolve potential conflicts—in essence, the method of analysis rather than the result is needed.

CONSEQUENCES OF PRIVATIZATION

Structure and coordination

Significant structural changes are occurring in agriculture—not only in size and ownership of farm firms but also in the linkages and level of coordination of farm production activities with input suppliers and product purchasers. More and more of these linkages are occurring through personally negotiated contractual or ownership arrangements rather than open markets. Although numerous forces and drivers are contributing to these structural changes, information and knowledge play a significant role. As in other industries characterized by contractual or ownership linkages, those individuals with unique and accurate information and knowledge have increasing power and control in the food production system. And with power and control is the capacity to garner profits from and transfer risk to others with less power.

The increasing role that knowledge and information play in obtaining control, increasing profits, and reducing risk is occurring for two fundamental reasons. First, the food business has become an increasingly sophisticated and complex business in contrast to producing commodities as in the past. This increased complexity means that those with more knowledge and information about the detailed processes as well as how to combine those processes in a total system (i.e. a food chain approach) will have a comparative advantage. The second development is the dramatic growth in knowledge of the chemical, biological, and physical processes involved in agricultural production. This vast expansion in knowledge and understanding means that those who can sort through that knowledge and put it to work in a practical context have a further comparative advantage. Thus, the role of knowledge and information in success in the agricultural industry is more important today than ever before.

The logical question for individuals in the food manufacturing chain is how to obtain access to this knowledge and information. Particularly for the independent producers in the farm sector, this knowledge and information is obtained from public sources as well as from external sources such as genetics companies, feed companies, building and equipment manufacturers, packers and processors, etc. In general, independent producers obtain knowledge and information from external sources in much the same fashion as they have sourced physical and financial resources and inputs. In contrast, contract- or ownership-coordinated systems of manufacture in which production, processing, and distribution are integrated obtain their knowledge and information from a combination of internal and external sources. Many of these firms or alliances of firms have internal research and development staffs to enhance their knowledge and information base. The knowledge they obtain is obviously proprietary and not shared outside the firm or alliance; it is a source of strategic competitive advantage.

Furthermore, the research and development activities in contract- or ownership-coordinated systems are more focused on total system efficiency and effectiveness rather than on only individual components of that system; it is focused on integrating the nutrition, genetics, building and equipment design, health program, marketing strategy, etc. rather than on the areas or topics separately. In addition to more effective research and development, such alliances or integrated firms have the capacity to implement technological breakthroughs more rapidly over a larger volume of output to obtain a larger volume of innovator's profits. In the case of a defective new technology or failed experiment, contract- or ownership-coordinated systems generally have more monitoring and control procedures in place and can consequently detect deteriorating performance earlier and make adjustments more quickly compared to a system with market coordination.

As knowledge and information becomes a more important source of strategic competitive advantage, those who have access to it will be more successful than those that do not have access. Given the declining public sector funding for research and development and knowledge and information dissemination, which has been the major source of information for independent producers, the expanded capacity of integrated systems to generate proprietary knowledge and technology and adapt it rapidly enables the participants in that system to more regularly capture and create innovator's profits while simultaneously increasing control and reducing risk. This provides a formidable advantage to the contract- or ownership-coordinated production system compared to the system of independent stages and decision making.

Contract/ownership coordination

Production agriculture in the past has focused primarily on commodity products with coordination through open spot markets. The increased specificity in raw material requirements combined with the potential for producing specific attributes in those raw materials is transforming part of the agricultural market to a differentiated product market rather than a commodity product market. The need for greater diversity, more exacting quality control, and flow control will tax the ability of spot markets to coordinate production and processing effectively. Open spot markets increasingly encounter difficulty in conveying the full message concerning attributes (quantity, quality, timing, etc.) of a product and characteristics (including services) of a transaction. Where open markets fail to achieve the needed coordination, other options such as contracts, integration, or joint ventures will be used.

Related to the difficulty of spot markets conveying the proper information is the speed of information flows and the rate of adoption with different coordination mechanisms. In general, contract- or ownership-coordination results in more rapid transmission of information between the various economic stages and consequently enhanced ability of the system to adjust to changing consumer demands, economic conditions, or technological improvements. The ability of the production and distribution system to be more responsive and adjust rapidly to changing conditions is increasingly important with the increased rate of change in economic and social systems worldwide.

This ability to respond quickly to changes in the economic climate is critical to maintaining profit margins as well as extracting innovator's profits. Likewise, quickly recognizing erroneous decisions and making appropriate adjustments and corrections are essential to survival and success. Market coordination of systems characterized by biological lags cannot respond to changing conditions as quickly as an integrated or contract-coordinated system. That is, the response at one stage can be initiated only after price signals the need for change; with little flexibility for adjustment during the growing and maturing processes, the change in quantity or quality is realized only after a full production cycle. By their nature, contract/ownership coordination systems require more frequent and direct communication between the decision makers at each stage on a wider variety of product and service characteristics than is typically possible with more traditional spot markets. Thus, the improved information flows and more rapid adoption and adjustment allow contract- or ownership-coordination systems to function more effectively in rapidly changing markets.

These arguments suggest that in traditional commodity markets where specific attributes are not demanded, supplies are fully adequate and can be obtained from various sources, and information flows between the various stages is minimal, traditional spot commodity markets can function quite effectively and efficiently. As one deviates from these conditions—which is increasingly the case with more specificity in raw materials, information flows, and fewer potential sources of acceptable supplies—various forms of contract/ownership/joint venture/strategic alliance coordination systems become more effective and necessary for efficient functioning of the production and distribution system.

The distribution channel

Changes in information technology along with the increased privatization of information and the expanding role that information has as a source of sustainable competitive advantage is having profound impacts on the agricultural input supply and distribution chain. An increasingly broad spectrum of information is flowing through the chain from supplier to end-user—information about product performance and characteristics, about customer demographics and buying behavior, and about support and service dimensions of the product and the customer relationship. The increased information demands of end-users require more informed and knowledgeable personnel throughout the entire intermediate stages of the distribution channel; customers or consumers don't want to do business with people who are technically less knowledgeable than they.

The speed of information transfer through electronic technology has shortened response time and smoothed product flow, thus reducing transactions costs in distribution channels. In some cases logistics and product flow is being separated from service, with product flow being more direct from manufacturer to consumer and service provided by a highly trained, informed local technical sales representative. In other cases highly trained, technical support personnel are accessible through 800 numbers, E-mail, FAX, and video telephone. As information becomes a more important source of sustainable competitive advantage in the future, the distribution chain is likely to include fewer participants, become more responsive, include tighter linkages and more interdependencies, and become harder to enter or replicate by firms not in the information loop.

Distributional consequences

Privatization of information has profound distributional implications. If the public role in providing data and information declines and private sector activity increases as is now occurring, there will be three distributional consequences. First, a major purpose of public information and data services has historically been to provide open access to potential users irrespective of size or other characteristics. Expanded private sector activity in the information markets would result in more of the information being provided to capture rents rather than for the common good. Thus information access will generally become less open (more restricted) as public sources decline in relative importance (or begin to exhibit rent-seeking behavior like private sector providers).

Secondly, with rent seeking behavior more important as a determinant of who is the target audience for private sector information providers, those who can and will pay the most rent will obtain the most and best information. One would expect that the largest, most sophisticated, and most specialized firms could pay higher rents and would receive more attention and emphasis by private sector information providers. Those not exhibiting these preferred characteristics would receive less attention.

Finally, private sector information providers would extract rents or capture value, thus redistributing revenues from the production sector to the service sector. Note however, that if the information increases efficiency and adds value, it has the potential to increase incomes of both the information provider and the producer, depending upon the rents extracted and how the incremental revenue is shared.

An important implication of the increased privatization of information for public sector information providers is that they will face even more questions concerning open access to their information, constraints on availability, targeted audiences, etc. (i.e., who gets the information and at what cost) because of growing concerns about the distributional consequences in terms of benefits and economic/political power of differential access to information.

Globalization of information

At the same time that information is becoming a more critical resource for success in production agriculture, globalization is fundamentally changing the nature of competition in the agricultural industry. During the 1970s and 1980s, two critical changes occurred in agricultural technology around the world: 1) both public sector and private sector investments increased in almost all geographic regions of the world, and 2) more technology and innovations were shared across national borders through public sector international research centers and internationalization of many agribusiness firms (Pardy 1991). This globalization of agricultural research and development in technology is a significant contributor to increased international competitiveness in agricultural product markets; no longer does one country or region of the world have exclusive and unique access to the latest information and technology with respect to genetics, nutrition, health management, and pest control.

In recent times the combination of globally adaptable production technology with site specific information on soils and climatic conditions that is increasingly available on a world-wide basis has added to the intensity of international competitiveness. Information as noted earlier is increasingly a source of competitive advantage, and information is now being acquired and transmitted globally. The significant and profound implications that follow are that internationalization of information and technology markets contributes further to international sourcing of products by agribusiness companies, international distribution of inputs by supply firms, and generally increased global competitiveness in the agricultural sector (Pray 1993). A logical and yet largely unresolved public policy challenge of this internationalization process is that of international intellectual property rights which we will discuss briefly later.

POLICY ISSUES

At this point we will identify a number of policy issues that merit debate and discussion with the increased privatization of information and technology transfer. Our approach will be to raise important questions, not provide definitive answers—that is the purpose of the public policy debate.

Role of the public sector

With increased privatization of information and an expanded role of the private sector in information markets, a logical public policy question is: What is the appropriate role of the public sector? One way to answer this question is to identify what the private sector will not do—what information it is not likely to provide—and then determine if there is a public need or benefit from having this information.

As suggested earlier, the private sector has no incentive to provide information if it cannot capture value or extract rents from doing so. A public benefit may accrue in the following three cases of market failure in information markets (numerous others might also be identified).

1. Some production firms may be too small or have other characteristics so as to not be desirable and profitable customers of private sector information providers, yet society may judge it inappropriate to discriminate against these firms in terms of access to information. In fact, this in part was one of the original justifications for the development of the land grant system of research, extension and education. An alternative public policy response would be to mandate private sector firms to service those producers irrespective of their profit potential as a condition of doing business, much like the Community Reinvestment Act requires lenders to provide financial services to their community, and anti-redlining rules that prohibit geographic discrimination in lending practices.

2. The private sector is not likely to provide objective information for public policy development and assessment. For many public policy issues, private sector firms would find it difficult to capture benefits of

providing information to make informed, unbiased judgments. For some issues, particularly related to regulation and legislation restricting their activity, private sector firms have an incentive to produce biased information into the public policy debate. In essence, the public sector plays a critical role in informing the decision makers and the public at large about the alternatives and consequences in the public policy arena.

3. As a logical extension of 1) and 2) above, the private sector is unlikely to provide information concerning public goods or common access resources such as the environment, wildlife, natural habitat, clean air, public recreational resources, etc. Efficient and effective management and allocation of these resources and utilization of public or common property goods will most likely require a significant public role.

Growing competition by private sector providers and decreasing budgets of public sector providers of information will require significant reassessment of the role of the public sector in the information markets. One way to proceed with this assessment would be to obtain concrete answers to the following fundamental questions: (1) Will the private sector provide the information needed? (2) If so, will that information have the access, objectivity, reliability, timeliness, and completeness that will meet public social and ethical standards as well as private profit incentives? With respect to information concerning public goods and common property, the public role is essential since private sector firms have no incentive to provide unbiased information unless they are regulated or legislated to do so.

One must recognize that even with unbiased information, the private sector may not allocate common property or public goods in a social welfare efficient manner because of the shorter time horizon and higher time preference for monetary benefits for private sector firms compared to the public interest. Consequently, additional public sector intervention beyond providing unbiased information may be necessary to obtain maximum social welfare from common property and public goods.

Public/private linkages

Producers can obtain information from public sources such as the land grant system and extension services, as well as community colleges, high school, and technical school adult education programs; or they can obtain information from the growing number of private sector providers. Private sector providers include those firms that provide information for a fee such as consulting companies, record keeping services, newsletters, and the rapidly growing number of electronic delivery systems and networks such as DTN and Farm Dayta as well as input supply firms such as seed and chemical companies that use information as part of their competitive strategy. In fact, more and more input suppliers are using information as a source of strategic competitive advantage. This information is not only in the form of advertising of product attributes, but is also broader in focus providing management assistance for example in the form of crop scouting, soil testing and mapping, nutrient analysis, and feed analysis.

The rapid growth in private sector information services and providers combined with the declining support for public sector information sources such as the extension services raises important questions concerning the potential linkages between public and private sector providers of information. The issue of the Extension Service becoming a wholesaler rather than a retailer of information is one such question. Should Extension train and provide information to private sector consultants, scouts, and management specialists who then have one-on-one contact with producers, and in this format provide more useful information to those producers because it is context and decision specific? Should Extension personnel train and provide information to technical representatives and sales personnel from input supply firms such as feed, seed, and chemical companies who will then provide this information to individual producers? Should the Extension Service utilize the private sector electronic media as well as newsletters and print media as part of their information dissemination/distribution channels? What are the risks and benefits to the public sector of these linkages? Will objectivity be compromised? Will producers who can't pay for the service be deprived of the latest information? Who will receive the recognition and rewards for providing the information? Will the taxpayer continue to support public sector information services if they don't have direct contact with it? What if the information is wrong—who will bear the risk of errors and possible liability? Without these private sector linkages, will public information services be able to be effective?

These are difficult questions to answer, but one possible way to at least approach an answer is to return to the concepts of knowledge, data, and information as defined earlier. As one thinks about the relative comparative advantage of the public sector vs. the private sector in this area, the public sector probably has a comparative advantage in analysis and integration—in essence, in knowledge. In contrast, the private sector probably has a comparative advantage in dissemination—in data gathering and manipulation and providing information. To be useful in decision making, knowledge must be integrated with data to generate information. One of the significant advantages of public/private sector linkages would be to allow each of the sectors to exploit their comparative advantage by combining the analysis and integration capacity of the public sector (the knowledge component) with the dissemination capacity of the private sector (the data and information component) to improve the information content and usefulness of the messages that producers receive.

Access to information

Because of the increased value of information and the expanding role of the private sector in providing it, the issue of the proprietary nature of and access to data and information becomes more important. With the increasing value of information and its use to obtain strategic competitive advantage, there is less free exchange of data and information. The issue of who owns the data and information becomes critical. For example, with respect to site specific soil characteristic information, who owns it—the operator who paid for it, the service company that gathered it, or the landowner who has title to the property? Can a farmer obtain this information from one company such as a fertilizer or chemical dealer and then provide it to a competitor who might have a lower price on fertilizer or chemical products? Does it make a difference if the farmer pays for the service and how much he pays, or if the information service is provided as part of a bundled package with the product? If coordinated production systems have the potential to obtain superior information as noted earlier, how can a producer that is not part of that system obtain access to similar information to remain competitive? Will you need to become part of the system—"in the loop"—to obtain access to the latest information to be competitive?

In a broader context, the public policy issue of intellectual property rights and the role of the public sector in making information a public good that is broadly available to all potential users becomes critical. The intellectual property rights debate has historically focused more on research and development and new innovations protectable under patent or copyright law. Particularly in agriculture, the public sector has played a major role in the research and development activity, and has thus provided broad access to new technology and ideas. In this context, part of the public purpose was developing and disseminating new ideas in a sufficiently broad fashion that a wide spectrum of users benefited and individual firms could not restrict access and capture the value associated with the new idea. Thus, one of the public sector roles was that of leveling the playing field so that all participants competed on the same grounds with access to new ideas and information. As more and more of the research and development (and thus new ideas) come from private sector firms compared to the public sector, and more of the information dissemination system becomes privatized, individual firms have more potential to capture value at the expense of end-users. They have the potential to restrict access to new ideas and information to particular users, thus favoring some producers and excluding others from the ideas, technology, or information necessary for them to be competitive.

Global intellectual property rights

Although significant progress was made in the negotiations leading to the General Agreement on Tariffs and Trade (GATT) on establishment and harmonization of international intellectual property rights, much progress remains on implementation with respect to intellectual property issues. Furthermore, some of the most flagrant violators of intellectual property rights are not signers of GATT or members of the World Trade Organization (WTO). As information and intellectual property increase as a determinant of international competitiveness, conflicts between countries will increase without further agreement. Increased privatization of information means less global sharing of new technology and innovation if international intellectual property rights are weak and pirating is common-place. The implications for growing disparities in economic growth and development by country and region are as significant as the distressing impacts on economic growth in the less-developed world.

The concepts of intellectual property rights including patent and copyright law as applied to agriculture were developed in an era of domestic markets and national firms, a relatively large public sector research, development, and information dissemination systems, and with a limited role of information as a critical resource. These concepts should be reevaluated in the current context of global markets and multi-national business firms, the shrinking role of the public sector in research and development and disseminating information, and the increasing importance of information compared to other resources as a source of strategic competitive advantage.

A FINAL COMMENT

The system and mechanism by which farmers obtain information to improve their decision making is changing dramatically. With increased context specificity and decision focus, information is becoming more valuable. Public sector providers of information have fewer resources. Private sector providers of information are expanding their activities and information has become a source of strategic competitive advantage for both farm input and service providers. Information is becoming an increasingly important driver of control and structural change in the agricultural industry, and information and intellectual property rights are becoming an increasing source of conflict and controversy as information increases in value and that value can be captured by private sector firms.

In this rapidly changing environment where knowledge and information resources have become increasingly valuable to consumers, producers, and society in general, the public sector knowledge/information system focused on agriculture in the form of the Extension Service and the land grant/USDA complex must reevaluate its role. This discussion has certainly not provided concrete answers to that new role, but by raising questions that will stimulate dialogue, it hopefully will contribute to a new vision of the public sector role in the agricultural industry.

REFERENCES

Drucker, P. 1992. For the Future: The 1990's and Beyond. New York: Dulton.

Ortmann, G., G.F. Patrick, W.N. Musser, and D.H. Doster. 1993. Use of private consultants and other sources of information by large cornbelt farmers. Agribusiness An International Journal, 9(4):391-402.

Pardy, P.G. 1991. Patterns of investments in agricultural research: New U.S. and international evidence. The Economic Impacts of Agricultural Research: A Workshop Sponsored by the Agricultural Research Institute, Washington DC: Agricultural Research Institute.

Pray, C. 1993. Trends in food and agricultural R&D: Signs of declining competitiveness? In U.S. Agricultural Research: Strategic Challenges and Options, R. Weaver (ed.). Agricultural Research Institute, Bethesda, Maryland, pp. 51-65.

Peters, T. 1992. Liberation Management: Necessary Disorganization for the Nanosecond Nineties. New York: Alfred A. Knopf.

3

Can Privatization of Information Meet the Goals of a Sustainable Agriculture?

Dennis Keeney and William T. Vorley

Society often is the victim of its near-term vision. Critical decisions that affect the long term are made on the basis of current trends, resulting in linear directions in policy that seldom are corrected until they prove ultimately unworkable. Sometimes forces converge that result in decisions that are critical to the long term direction of an industry or a society. In June 1992, those forces converged at Rio de Janeiro for the UNCED "Earth Summit". Ambitious roles of government were laid out in the action plan for sustainable development known as Agenda 21. Chapter 14 of Agenda 21, which covers agriculture and rural development, makes numerous recommendations for "governments" to act for building sustainability in agriculture. Importantly, no demands are made of the private sector. This assumption, that government action will deliver the goods of sustainability, now looks unconvincing in a new era of privatization and the reassessment of the role of government. In the name of enhanced efficiency, effectiveness, and accountability, governments are putting greater reliance on markets to deliver agricultural services (Carney 1995). As governments step back from agricultural extension, who will represent the larger interests of society, and the interests of future generations, for the farmer? Who will develop and transfer information for the protection of natural resources and for the enhancement of social capital? Who will redesign agriculture for long term sustainability?

ISBN 1-57444-104-3/98/$0.00/$.50

SUSTAINABILITY: A PUBLIC RIGHT, A PRIVATE INTEREST, OR BOTH?

A sustainable agriculture is one that maintains the natural resource base while providing food and fiber for the world's population for the indefinite future. This will require an agriculture that will adapt to modifications in climate, population trends, markets, environment and energy trends. Information on new (and rediscovered) technologies and farming systems to meet these needs will be required at an ever-increasing rate.

Moving to sustainable agriculture systems will involve public research and extension as well as agribusiness in the development of sustainable technologies and transferring the information to the operator, and will involve policies that encourage development of new farming strategies. But in this era of government downsizing and economic restructuring, it is naive to expect a reversal in the trends towards diminished size and field presence of public extension services. Farmers will increasingly rely on service from the commercial sector and from independent crop consultants.

The prominent role for the private sector has caused concern among advocates of sustainable agriculture, who fear that the chicken coop of sustainability is being left to the foxes of profit. The profit motive implies that private information must be in the commercial interest of the provider as well as the recipient.

Private industry's belated moves to embrace sustainable development as a post-Rio touchstone have been largely on its own terms, which critics identify as far along the continuum of values towards productionism and technological optimism. This is clear even from the choice of definitions for sustainability. For instance, DowElanco have written that "sustainable farmers optimize inputs (financial investments) to maximize profits" (DowElanco 1994) and BASF have defined sustainable agriculture as "that level of productivity that allows the agricultural enterprise to be economically competitive" (Richgels 1990). The pesticide industry in particular does not hesitate to justify its current mission with the opinions of Dennis Avery (1995) and Paul Waggoner (1994) (see Johnen 1995).

As society's priorities have shifted from maximizing output to optimizing inputs, companies are making big efforts to improve the cost-benefit ratio of intensive agriculture through rate reduction, precision agriculture, farmer training and development of new "soft" technologies. This puts private industry firmly in what MacRae et al. (1993) categorize as the "efficiency" and "substitution" stages towards agricultural sustainability:

> "In the *efficiency* stage, conventional systems are altered to reduce both resource waste and environmental impact, e.g., by banding fertilizers and monitoring pests.
>
> In the *substitution* stage, finite and environmentally disruptive products are replaced by those that are more environmentally benign, e.g., synthetic nitrogen sources by organic sources and non-specific pesticides by biological controls."
>
> The efficiency and substitution stages, by producing "more with less,"

offer clear benefits in terms of producing more with fewer chemicals and less land.

While it can be forcefully argued (Avery 1996) that input-intensive agriculture will lower the impact of agriculture on natural resources and wildlife, the wise and efficient use of inputs may represent the perfection of means and confusion of ends. For when we talk "sustainability" we must ultimately ask what is being sustained, for how long, for whose benefit, at what cost, over what area, and by what measurement criteria. MacRae et al. (1993) note that because "neither of efficiency nor substitution strategies confronts the causes of problems, producers remain reliant on externally derived curative solutions and inputs." They introduce the "redesign" stage to address the future:

> "The *redesign* stage seeks to avoid problems by design and management approaches that are appropriate in terms of their physical, temporal and socio-economic context." (MacRae et al. 1993)

An excellent example is the attempt to lessen herbicides in surface water. An efficiency response would reduce rates by improved timing and placement, perhaps by precision farming, aiming at weed suppression rather than clean fields. Substitution would replace high rate, water soluble products with a new class of immobile, rapidly degrading chemicals, or even substitute herbicides with smother crops. In contrast, redesign thinking would ask why herbicides are being used in the first place, and could review the arable production of grain crops for livestock feed, resulting in a switch to grazing. An intermediate step towards redesign might be a landscape approach where filter strips would be placed along banks of creeks and streams to filter out agrichemicals and silt. Clearly the efficiency, substitution and redesign (ESR) concepts overlap and can not be delineated clearly.

"Redesign" is virtually an unknown territory to the private sector. Corporate environmentalism has been reduced to managerialism (Sachs 1992). An obsession with reduced emissions and regulatory compliance imposes a rigid order on a system, and prevents the kind of redesign that can allow a system to evolve. By relinquishing agricultural information to for-profit enterprises, we risk locking agriculture into a tail-chasing circle of incremental improvements of efficiency and precision, which excludes precautionary language of "redesign," with its emphasis on the long term sustainability and health of food, ecosystems and communities.

Regrettably we must acknowledge that the public sector has not done much better than the private sector in promoting the shift towards regenerative agriculture. The era of publicly-supported research and extension in agriculture has been one of enormous gains in production and productivity, but also one of unrelenting structural change, displacement of farmers and farm workers, environmental degradation, and declining public confidence in food quality. Technological developments originating largely with public sponsored research and education have constantly favored this trend (Buttel 1996; Urban 1996). Thus, when current agricultural practices are labeled "unsustainable," mainstream public institutions and agribusiness are regarded interchangeably as "the system." The reason for the lack of redesign emphasis lies in the shared paradigm of technological optimism by scientists and administrators in both the corporate and academic research establishments (Thompson 1995). Critics find it much easier to blame the pervasive corporate influence and sponsorship of research. While only a small portion of land grant agricultural research is directly sponsored by industry, the funds often represent much of the discretionary money available and thus corporate sponsorship can have a disproportionally high influence on the direction of the research. However, this need not be the case if priorities and rewards emphasize sustainability rather than publish or perish.

Most of the university and federal government-based sustainable agriculture programs originated as responses to outside criticism. They are a delayed response to the recognition of society's wider expectations from agriculture beyond the simple production of commodities. They include the provision of landscape and recreational amenities and the protection of water and air quality. Proponents of redesigned agriculture, such as Wes Jackson, Marty Strange, and Alan Savory, have felt compelled to set up independent institutions for research, extension and advocacy outside of the university system (e.g., Jackson 1990). However, few of the more holistic research and outreach programs, be they public or non-government organizations, have gained the preeminence of the established programs, and most continually struggle for sufficient funding to make a difference.

We should also acknowledge that many commentators do not accept that the private sector is incapable of redesigning itself to promote and profit from the redesign of agriculture, transport, architecture, or a host of other societal challenges. It is interesting to note that the ESR continuum is shaping not only the debate about agricultural sustainability, but also the discussions over the greening of business. Business leaders, especially from the chemical industry, have publicly endorsed the concept of sustainable development as "good business sense because it can create competitive advantages and new opportunities" (Carson and Moulden 1991; Schmidheiny and BCSD 1992). But there exist large differences concerning what those "new opportunities" may be; from simple efficiencies in production to profound "biorealistic" redesign of a corporations' goals and products (Hawken 1993).

So the stereotypes of public research and extension for the public good vs. private research and extension for profit at all costs are not a realistic portrayal of the past, nor by implication, an accurate reference for the future. The debate is not so much between "public" and "private," but more between productionism, which attempts to achieve sustainability through efficient use of safer inputs, and holism, whereby sustainability is sought through redesigning agriculture along ethical and ecological principles.

In this paper, we use the ESR for both agriculture and industry, to ask what the private sector has done in terms of information for sustainable agriculture, what it could do, and what the implications are for the structure of business and the role of its regulators. We review each stage of the ESR model and describe for each the role of the source of information. A case study is used to illustrate the challenges of enlisting private sector capital in the redesign of agriculture.

EFFICIENCY, SUBSTITUTION AND REDESIGN IN AGRICULTURE AND INDUSTRY

Efficiency

True efficiency is not just about getting more benefits from fewer inputs, be they chemicals, energy, labor, or land. It is about getting prices right so that farmers and their customers are confronted with the full cost of their actions (Hawken 1993; Schmidheiny & BCSD 1992). If markets are corrected so that the prices of chemical inputs account for external environmental and social costs, the private information market would support a sustainable agriculture. But until society acts to place externalities in the economic equation, goods with high external costs will be oversupplied.

Industry has embraced the efficiency revolution in search of the elusive win-win-win: pollution prevention, cost reduction and environmental public relations. Emphasis is placed not on what is being produced, but how a product is made, with the aim of reducing material throughput and waste. This approach by and large has produced the desired short term results but falls far short of correcting problems in the long term.

A similar pollution prevention philosophy is logically being applied to agriculture. Technologies such as improved chemical products, better tillage and application technologies, and precision farming fit well the corporate culture and business (MacRae et al. 1993). But even with new information technologies, efficiency information is complex and site-specific. It requires more rather than less operator intervention just when the agricultural sector is being depopulated and the ability of cooperative extension to supply information is in decline. Farmers vary in their resources and interest in collecting and processing information. Many have sufficient resources to make management decisions on information developed internally, while another source of advice may be specialists ranging from the input supplier complex to food processors or retailers who are becoming more involved in farm-level decision making. The major groups emerging in the private agricultural information market are crop consultants, either independent or tied to a cooperative or agrichemical dealership. There are predictions of the demise of independent retail dealers as marketing of agrichemicals and seeds move to the same mass marketing techniques used in the urban consumer markets (e.g., Wal-Mart outlets). Information on the use of chemicals would be provided by the manufacturer/formulator, eliminating one or two intermediate handlers of information and lowering transaction costs.

There has been considerable concern at the prospect of the extension vacuum being filled by private farm advisors with the potential for conflicts of interest. The Certified Crop Advisor (CCA) program administered nationally by the American Society of Agronomy (ASA) has attempted to address this issue. This program developed in the early 1990's with leadership from ASA, industry and government agencies (Extension, NRCS, and EPA). It has expanded rapidly and now covers most of the nation. A key component of the CCA program is the ethics statement that must be signed before certification is issued. Another critical part of the program is the continuing education requirements to maintain certification (Nelson 1996).

But we propose that even independent crop consultants, whose only "product" is information, are also largely "efficiency" oriented. Farmers contract with a consultant firm to: (1) reduce (or optimize) their inputs and maximize their yields, (2) conduct the tedious and specialized work of crop scouting, soil sampling, and record keeping, and (3) provide a second opinion from the one provided by agrichemical dealers. There is no thought or resources given to redesign, and substitution is used only if economically expedient. The market would not seem to be able to support the more holistic approach as it is currently structured.

As the value of information at the farm level increases, either as a result of improved precision, improved quality, or price increases of chemical inputs, big business will start looking for ways to capture that value; perhaps as franchised service centers linked to a centralized cutting edge information and training network, or in the form of performance contracting modeled on Health Maintenance Organizations (HMOs). The entire decision making process would be removed from the land owner-operator. This marks a shift from selling products to selling solutions, with much more emphasis on the knowledge ingredient. It recognizes that information is the ultimate "clean" or "dematerialized" business, with almost no material throughput, no waste, no clean-up costs, and few liabilities. There is no reason why such service centers should not expand to offer services in agroforestry, planning, construction and maintenance of riparian buffer strips, manure crediting and management. If big business moves into the crop consultancy business, which has traditionally been the realm of the local entrepreneur, there are concerns that another chance for grass roots rural economic development will be lost. The economic benefits of precision agriculture and privatized crop consultancy are likely to be uneven, as these technologies and services are not size-neutral.

We should note that some market trends in the agrichemical industry are ambiguous in terms of efficiency. Agrichemical dealers and cooperatives are responding to intense competition by providing more information services, such as scouting or soil testing, "free" with the product or well below market rates. Such "free" service depresses the professional crop consultancy market by reducing farmers' price expectations. Competition between pesticide manufacturers has forced them to be more aggressive in backing products with performance guarantees and resprays so that farmers are less inclined to experiment with below label rates or to use information-intensive strategies such as scouting and precision weed management.

As the influence of food processors and retailers in farm decision making increases through contract or "preserved identity" production, farmers are locked into a prescribed set of crop production and protection practices which prevent them from moving beyond efficiency or substitution.

Substitution

Private research and development has been increasingly more active in substitution technologies because new seeds, "soft" chemicals, and biocontrol agents are patentable and commercially viable (Duvick 1995; 1996). But substituting valuable new products in an industrial farming system may not be sustainable.

For example, biological insecticides such as Bt (*Bacillus thuringiensis* endotoxin) may be lost to pest resistance just like chemical insecticides if used without consideration of basic ecological principles (Wagge 1995). In the end, substitution technologies will look much like efficiency technologies, with their own sets of advantages and drawbacks but they do offer industry the opportunity to come up with far more environmentally benign products.

Redesign

Compared with "efficiency" and "substitution," "redesigned" agriculture may not be perennially information-intensive (see Lockeretz 1991). Preliminary training (e.g., in holistic resource management) may allow farmers to proceed relatively autonomously (or supported by a network of similarly trained individuals) without information as a "purchased input." After all, many of the "inputs" are already there (e.g., natural enemies, nutrient cycles, rotations, and plant combinations) that suppress weeds and pathogens. Thus the purchase of skills is necessary only to conserve or intensify those indigenous resources for biologically-intensive agriculture, or as a last resort, supplement them with biorational products. Furthermore, a goal of many farmers is reduced dependency on externally-purchased inputs, whether they be chemicals or consultants. A survey of Nebraska farmers showed that use of consultants for crop production information declined from 37 percent with the "least sustainable" group to 0-8 percent with the "most sustainable" clusters (Bernhardt and Allen 1994). Farmers such as the Practical Farmers of Iowa are banding together to do their own information and demonstration programs. Fred Kirschenmann (1991) writes:

> "There is a widespread image held by society that farmers don't have the skills or imagination to do an effective job of farming, and are therefore dependent on outside professionals. And this is despite the fact that some of agriculture's biggest problems stem from the fact that farmers often followed the advice of such professionals."

In industry, a transition to redesign will mean companies challenging their reason for being or their core competencies (Bavaria 1994). For instance, is a chemical company in the pesticide business, the crop protection business, or the agriculture business? Hawken (1993) comments that

> "Companies must re-envision and re-imagine themselves as cyclical corporations... If DuPont, Monsanto, and Dow believe they are in the synthetic chemical production business, and cannot change this belief, they

> and we are in trouble. If they believe they are in business to serve people, to help solve problems, to use and employ the ingenuity of their workers to improve the lives of people around them by learning from the nature that gives us life, we have a chance."

There are strong internal and external barriers to the re-envisioning of a corporation, especially from shareholders. Institutional pressures force employees to suppress deeply-held ecological values (Glasser et al. 1994), and through intellectual inbreeding, corporations become entrapped in their own world view (Vorley and Keeney 1995). But as the farm input sector becomes industrialized, and as power and profits continue their slide from the farm toward processors and retailers of food, there are also strong reasons for big business to look carefully for opportunities in redesigned agriculture (Vorley et al. 1995).

The following case study traces the emergence of new concepts—integrated crop and pest management and precision farming, not as a success story but as an example of how the tension between efficiency, substitution and redesign approaches can have profound effects on the nature of the information which farmers receive.

INTEGRATED CROP MANAGEMENT (ICM) AND PRECISION FARMING

Integrated crop management and its components, nutrient and pest management, have been the domain of public research and extension, arising from concerns over environmental pollution from pesticide and fertilizer misuse. Private industry, especially the pesticide industry, was apathetic, or at best ambivalent, about ICM. But in recent years, ICM has caught industry's attention for several reasons. First, nutrient and pest management are—as integrated pesticide and fertilizer management—excellent product management tools for chemicals, to minimize risks (and prolong the commercial viability) of old products, and to develop niches for new selective chemistry. Secondly, farm equipment and aerospace and defense industries have seen considerable profit potential in packaging ICM with yield monitoring and global positioning and mapping technologies under the generic term precision farming (Vanden Heuvel 1996).

But this acceptance of ICM within the private sector inevitably stresses only those ingredients linked to product sales, which in turn distorts the concepts away from their ecological foundation or redesign potential. Integrated pest management (IPM) tends toward an economically rational use of pesticides rather than a reduced reliance on pesticides. Precision farming—which could be an extremely powerful tool for redesign—becomes axiomatic with the precise application of fertilizer or herbicides.

CONCLUDING REMARKS

The profit motive has brought us a highly productive agriculture. There is still much room for improvements in efficiency in nutrient and pesticide use; private sources of information have a vital role in implementing these improvements at the farm level. The ethical implications of private corporations consolidating crop production and protection information in order to package and extract value must be carefully studied despite its benefit for corporate greenness.

The biggest need appears to be research and information for the redesign of agriculture. This is not to imply that research and development on greater efficiency should not be continued. Established technology must be continually refined as new technologies come on line, as products are developed, or as new pests adapt to current farming systems. But sustainability is built on diversity, which in turn demands a range of viable agricultures as alternatives to a single, optimized status quo. Publicly funded land grant university research and extension should be focused in this area rather than duplicating or serving areas in which the private sector has advantages in research skills and field presence. It is difficult to envision private sources providing sustainability information considering current policies and market signals. Before the privatization process becomes a done deal, there needs to be a serious debate about the ability of a privatized farm information sector to deliver the goods on sustainability. Paul Hawken (1993) concludes that current commerce and sustainability are "antithetical by design, not by intention." He sees a design problem running through all business and in the institutions that surround it (Hawken 1993). Once the straw man arguments and vested interests of the sustainability debate are cast aside, there really may be a new prospect for a redesigned private sector providing information for redesigning agriculture.

REFERENCES

Avery, D. T. 1995. Saving the Planet with Pesticides and Plastic: The Environmental Triumph of High-yield Farming. Indianapolis, Indiana: The Hudson Institute. Herman Kahn Center.

Avery, D. 1996. Technology and agriculture: Empowerment or entrapment? Speech given before the National Agricultural Forum, Des Moines, Iowa.

Bavaria, J. L. 1994. The challenge of being green. Harvard Business Review. July-August, p. 40.

Bernhardt, K., and Allen, J.C. 1994. Adoption/diffusion of sustainable agricultural practices: What influences change? Paper presented at the Annual Rural Sociological Society Meeting, Portland OR August 10-14, 1994.

Buttel, F. H. 1996. The impact of changes in technologies on the structure of agriculture. Speech given before the National Agricultural Forum, Des Moines, Iowa.

Carney, D. 1995. The changing public role in services to agriculture: A framework for analysis. Food Policy 20:521-528.

Carson, P., and J. Moulden. 1991. Green is Gold: Business Talking to Business About the Environmental Revolution. Toronto: Harper Collins Publishers.

DowElanco. 1994. The bottom line on agri-chemical issues. Sustainable agriculture feature. Indianapolis IN.

Duvick, D. N. 1995. Biotechnology is compatible with sustainable agriculture. Journal of Agricultural and Environmental Ethics 8:112-125.

Duvick, D. N.. 1996. Seed company perspectives. In: Herbicide Resistant Crops. Agricultural, Environmental, Economic, Regulatory, and Technical Aspects. S. O. Duke (ed.). New York: CRC Lewis Publishers.

Glasser, H., P.P. Craig, and W. Kempton. 1994. Ethics and values in environmental policy: The said and the UNCED. In Toward Sustainable Development (ed. J.C.J.M. van den Bergh et al.) Washington D.C.: Island Press.

Hawken, P. 1993. The Ecology of Commerce: A Declaration of Sustainability. New York, NY: Harper Collins Publishers.

Jackson, W. 1990. Making sustainable agriculture work. pp. 132-141. In Our Sustainable Table. (ed. R. Clark). Berkeley, CA: North Point Press.

Johnen, B.G. 1995. Risk assessment and crop protection: Products use reduction. Paper presented at a workshop on pesticides, August 24-27. Wageningen Agricultural University, Netherlands, as part of the European Union Concerted Action Program Policy measures to control environmental impacts from agriculture.

Kirschenmann, F. 1991. Imagination—a must for sustainable farming. New Farm Sept/Oct. pp. 45-48.

Lockeretz, W. 1991. Information requirements for reduced-chemical production methods. American Journal of Alternative Agriculture 6: 97-103.

MacRae, R.J., J. Henning, and S. B. Hall. 1993. Strategies to overcome barriers to the development of sustainable agriculture in Canada: The role of agribusiness. Journal Agricultural and Environmental Ethics 6:21-51.

Nelson, J. 1996. Reaching thousands. Insider. Ag Consulant. Jan 1996. p. 9.

Richgels, C.E. 1990. *Quoted in:* Sustainable agriculture: Perspectives from industry. Journal of Soil & Water Conservation, Jan-Feb 1990, 31-33.

Sachs, W. 1993. Global ecology and the shadow of 'development'. Pp 3-21 in Global Ecology: A new arena of political conflict. Ed. W. Sachs. Zed Books, London & New Jersey.

Schmidheiny, S., and the Business Council for Sustainable Development. 1992. Changing Course: A Global Business Perspective on Development and the Environment. Cambridge, MA: MIT Press.

Thompson, P. B. 1995. The Spirit of the Soil: Agriculture and Environmental Ethics. London & New York: Routledge.

Urban, T. N. 1996. Acceptance speech for the Agricultural Vision Award. National Agricultural Forum, Des Moines, Iowa. March 4, 1996.

Vanden Heuvel, R. M. 1996. The promise of precision agriculture. Journal of Soil Water Conservation 51:38-40.

Vorley, W.T., and Keeney, D.R. 1995. Sustainable pest management and the learning organization. Paper presented at the International Food Policy Research Institute workshop Pest Management, Food Security, and the Environment: The Future to 2020. Washington DC: May 10-11 1995.

Vorley, W. T., D. R. Keeney, D. Koechlin, M. M. Mayhew, E., Ozdemiroglu, D. W. Pearce, J. N. Pretty, and R. Tinch. 1995. Can the Pesticide Industry Benefit from Sustainable Agriculture? In Press.

Wagge, J. 1995. Divergent perspectives on the future of IPM. Paper presented at the International Food Policy Research Institute workshop Pest Management, Food Security, and the Environment: The Future to 2020. Washington DC: May 10-11.

Waggoner, P. 1994. How much land can 10 billion people spare for nature? CAST Task Force Report 121. Feb. 1994.

4

Integrating Public and Private R&D in the Food and Agriculture Sector

Don Holt

The issue of whether public food and agricultural research and education programs should be privatized needs to be analyzed in the context of changes now underway in agriculture. These changes are toward greater specialization, more specialists, greater division of labor, more complex value-added processes involving more intermediate products and markets, more highly coordinated markets, and more highly coordinated research and development (R&D) programs. Currently, public and private food and agricultural research, education, and outreach are inextricably intermingled and generally must be that way to function effectively. Seed-related research and outreach are particularly good examples of this complexity.

It is important to distinguish at least two facets of the privatization issue, namely who should conduct research and who should pay for it. It is also important to distinguish between the traditional extension programs that some people would like to privatize and other kinds of university outreach. Many aspects of outreach cannot be privatized, because they consist of complex relationships among researchers, teachers, extension educators, outside intermediaries, and/or constituents, some of whom are already in the private sector.

Early adopters of new technology reap benefits that are soon competed away when the technology is more widely adopted. Most of the benefits of publicly funded agricultural R&D accrue to the public as consumers, so it is appropriate

I wish to acknowledge the valuable help of Kathleen "Casey" Drury in gathering information and reviewing literature for this paper.

ISBN 1-57444-104-3/98/$0.00/$.50

for the public to pay most of the cost. In fact, a major effect of publicly supported R&D has been to foster competition among producers, thus leading to higher quality, safer, more affordable and convenient agricultural products.

In this paper, the case is made that producers, broadly defined, cannot afford to pay much of the cost of R&D that does not provide them proprietary benefits, i.e., benefits that are not available to competing producers. Likewise, private firms will not expend their R&D resources on programs that do not yield proprietary advantages. Since there are many beneficial agricultural research and education programs that do not provide competitive advantage for farmers or private firms, there will continue to be appropriate roles for both the public and private sectors in food and agriculture research.

By better communication and coordination between universities and their food and agricultural constituents, the function of the market for research and education services can be improved. Some experiences in Illinois reinforce this contention.

BACKGROUND

The issue of privatization needs to be analyzed in the context of important trends in the food and agriculture sector. The most important trend is that the sector, once populated almost entirely by farmers, continues to develop new specialties, involve more specialists, and undergo further division of labor (Holt 1993). Markets for consumer products and services are becoming more complex, involving more intermediate markets and economic stages, and requiring more vertical coordination (Barry et al. 1992).

Likewise, the process of developing and transferring new technology is becoming more complex (Holt and Sonka 1995). For example, to improve the nutritional quality of a meat product might require changes in several stages in a complex value chain involving crop germplasm, crop seed, grain, feed, breeding livestock, growing and finishing livestock, meat packing, meat processing, meat distribution, meat preparation and retailing, and meat consumption and utilization. Changes in converging value chains (those that provide inputs to meat production and utilization) and diverging value chains (those that utilize by-products of meat production and utilization) might be required to make the overall change economically viable.

R&D projects and programs leading to such changes will of necessity be similarly complex. They will need to be integrated over several disciplines, R&D functions (basic, developmental, and adaptive research and technology transfer) (Holt 1991), and stages in complex value chains. It will probably be necessary to address cross-cutting environmental, social, legal, regulatory, and/or logistical issues which involve additional disciplines. Also, there are stages in R&D efforts. Important decisions should be made between these stages, including decisions on team composition and "go/no-go" decisions. This phasing adds further complexity (Cooper 1990).

It is unlikely that any individual university or private firm has all the disciplines, specialists, facilities, equipment and other resources, and relationships required to mount all activities involved in such complex, outcome-oriented, R&D efforts. Inevitably, many such efforts will require interdisciplinary, interinstitutional, and interorganizational teams and extraordinary levels of communication, coordination, and integration. I believe this need is driving the increased interest in outcome-oriented planning, e.g., the Government Performance and Results Act of 1993 (GPRA) (Chief Financial Officers Council 1995). GPRA is federal legislation requiring that federally funded programs and projects include a strategic plan that specifies goals stated in terms of desired practical outcomes. Programs and projects will be evaluated in terms of whether or not the goals (outcomes) are actually achieved.

THE ISSUE OF PRIVATIZATION

Recent unprecedented public scrutiny of land grant institutions focuses on several issues, one of which is whether public funds should be used to support agricultural research and extension. Several privatization approaches are proposed and advocated. But what constitutes privatization? Does the issue concern who should conduct the research or pay for it?

At one end of the spectrum of views, there are those who believe that most if not all food and agriculture research and extension could and should be carried out by private firms without government involvement. Proponents of this approach point out that private sector investment in agricultural research has grown dramatically relative to public sector investment. It is now larger by a factor of 1.67 (Huffman and Evenson 1993). Some contend that private firms and consultants already replaced extension personnel as primary sources of information for farmers.

In general, proponents of privatization believe that private firms, disciplined by the bottom line and by customers with long memories, conduct R&D more efficiently than public institutions. They argue that the historic public food and agriculture research institutions including state agricultural experiment stations (SAES) (Holt 1994) and the Cooperative Extension Service (CES), were useful at one time but are no longer needed.

In a variation of the privatization theme, government agencies and institutions would contract with private firms to conduct agricultural research and extension.

Presumably, the contracts would specify the subject matter and geographical scope of research, describe expectations for precision and accuracy of results, and provide for dissemination of the resulting information. As in the case of complete privatization, SAES and CES would no longer be needed. In still another privatization approach, public agricultural research and extension programs would be retained but all or part of the cost would be recovered through user fees.

At the other end of the spectrum are groups that believe that strong public research and extension programs are still needed to address issues of broad societal concern, such as environmental and sustainability issues, social issues, food safety, and preservation of rural institutions, including the family farm. These groups fear that the private sector will not elect to conduct or pay for research and educational programs that do not contribute to profits. They would not favor complete or even increased privatization of agricultural research and extension. My impression is that most university and other public sector scientists and administrators would not favor legislative action to privatize publicly funded food and agriculture R&D. They probably have both selfish and unselfish reasons for taking this position. Some legislators and taxpayer advocacy groups advocate privatization, because they see this as a way to cut government costs. I know few if any private sector managers, even R&D managers, who favor complete privatization of food and agriculture research.

COMPLEX DIMENSIONS OF THE PRIVATIZATION ISSUE

To appreciate the complexities of this issue, it is important to realize that since 1969 private firms have considerably increased their investment in SAES research, such that now they provide about 20 percent of that support (Pray 1993). Also, because of the creation of commodity checkoff programs and the inclination of checkoff boards to invest in public sector research, commodity organization support of SAES research and extension education programs increased dramatically since 1980. For example, soybean checkoff programs did not exist before 1975; now a national soybean checkoff program generates between $20 and 40 million annually that is spent on research, education, and promotion projects.

University/industry research relationships exist in several forms, including faculty consulting, industrial affiliates programs, research consortia, research centers, research partnerships, university research parks, and formation of spin-off corporations (Jarvis et al. 1994). These relationships are generally fostered and supported by state and federal legislators who believe they speed up the process of translating scientific findings into practical knowledge and technology. Universities foster these relationships because they attract public interest and private investment in research programs and provide valuable educational experiences and contacts for students.

Should these programs be considered private sector or public sector research? Is this evidence that at least some private firms and organizations regard public sector research as a good investment? Do they do this because some of the costs, especially fixed costs, are paid from public funds? Would private firms conduct this research themselves if there were no SAES, USDA-ARS, or CES?

It is common practice for firms that sell agricultural inputs, such as seed, fertilizer, and chemicals to hold open meetings to promote their products. Often they will invite university extension specialists to make presentations concerning the results of research involving their products and about production practices in general. University scientists are often involved in research on rates, dates, and methods of application of various inputs and research on which label specifications are based. Thus, they have valuable information to share with the consumers of agricultural inputs.

These same firms voluntarily enter their products in tests, such as variety trials, herbicide and insecticide comparisons, and fertilizer experiments conducted by universities. They do this knowing that in these trials their products will be compared with competing products and may not perform as well. These trials are definitely not basic research and yet most farmers and agribusiness people regard such comparisons as appropriate activities for public sector scientists. They look on the public institutions as unbiased sources of information.

Private firms frequently contract with universities to conduct research, including basic research, with the understanding that if they fund the research, they will get proprietary access, through exclusive or semi-exclusive licenses, to the resulting new technology. Universities almost always retain ownership of intellectual property they generate. They license the use of this property. The licenses may be exclusive or non-exclusive and may or may not require royalty payments, which universities use to offset research costs.

Often the information or materials generated in the basic research is not commercially useful without further R&D. In this situation, close cooperation and coordination between private and public entities are required to bring about the desired commercial result.

Universities are increasingly protecting intellectual property (inventions and discoveries) through patents, copyrights, and trade secrets and seeking to capitalize on it through royalty-bearing licenses. Most universities did not develop significant intellectual property administration capacity or policy until after 1970 (Matkin 1992). To illustrate the rate of change, twenty nine major universities were issued 535 patents in 1988, that number steadily increasing to 861 in 1992 (data supplied by Kathleen Terry of the State University of New York at Buffalo).

The attitudes of universities and some of their constituents toward intellectual property has changed. Traditionally, the philosophy was that, since university programs were supported by tax revenues, the resulting intellectual property should be freely available to members of the general public. More recently, university-generated intellectual property is seen as valuable public property, which the university is obligated to manage in the best interests of the public.

The public interest is usually best served by successful commercialization of useful inventions and discoveries. Thus at least some publicly owned intellectual property should be managed to maximize the likelihood of successful commercialization. Often that likelihood is maximized by exclusive licenses, which provide private firms incentive for further R&D toward commercialization.

Private sector support of university research provides fellowships, scholarships, and assistantships, helps offset the costs of undergraduate and graduate education, and saves money for governments, private firms, and individuals. It is hard to imagine how a university could train graduate students without mounting research programs. In fact, a major reason universities conduct research is to educate the scientists of the future in the philosophy and methods of research.

It is interesting to ponder whether extension programs might be privatized while related research programs remained public. Authors of other papers in this volume describe attempts by various nations to privatize extension (see Bunney, Cary, and Scarsbrick). One can envision taking the extension function out of a public institution, as long as that function is clearly a separate activity that can be carried out by different people rather than by those conducting research. It is much harder to envision taking the outreach function out of a university.

Outreach is accomplished in many different ways, one being the traditional Cooperative Extension approach. The other outreach mechanisms almost always involve relationships between university researchers and end users or intermediaries outside the institutions. The relationship of a researcher to decision-makers within a private firm, regulatory agency, commodity group, farm or trade organization, professional society, publishing house, lobbying firm, or other outside entity is often the conduit through which the technology and information generated by the researcher is implemented by the practitioner. The path is often tortuous, involving may intermediaries, both public and private. How does one "privatize" these relationships, which are often already partly private? My experience is that the outreach function is different and involves different proportions of private and public activity in each individual R&D project. In many R&D projects, the outreach function is inseparable from the research function and its private and public components are likewise inseparable.

These examples illustrate synergistic and almost inextricable intermingling of public and private interests in food and agriculture research and extension. I believe the system, the market if you will, naturally moves toward a dynamic and economically appropriate equilibrium between public and private programs and resources. A good example of how the public/private system adjusts to economic realities is research on genetic improvement of crop varieties.

PUBLIC/PRIVATE RELATIONSHIPS IN SEED RESEARCH AND EDUCATION

With the advent of hybridization and other improved plant breeding techniques pioneered in land grant universities, plant breeding matured as a science and started to make extremely valuable contributions to practical agriculture. Early in this process of change, almost all genetically improved crop varieties were developed and the seed multiplied by university agricultural scientists working with private, non-profit foundation seed and crop improvement organizations. Now, private firms produce almost all corn varieties and most varieties of horticultural crops; both private firms and universities produce soybean and fruit crop varieties, and universities produce most of the small grain varieties.

Some private firms have their own plant breeding programs. Some of these are larger than the programs mounted by individual universities. Smaller firms obtain crop varieties from universities or private variety brokers, in each case paying royalties or R&D fees. In some situations, the universities are in more or less direct competition with private seed firms. However, the same seed firms with which universities compete use university germplasm and genetic stocks in some parts of their programs.

The general trend is toward more private sector participation, but not necessarily less public sector participation. Based on trends, it is predictable that universities will continue to be substantially involved in basic plant breeding and genetics research, biotechnology, and germplasm enhancement. Some universities will continue to have a variety of development programs, in part to train graduate students for careers in plant breeding and in part to serve the needs of seed firms that are too small to have their own variety development programs.

There is a complex intermingling of public and private resources in commercial plant breeding. There are complex reasons for this division of labor. They have to do with whether crops are naturally self-pollinated, whether farmers must buy new seed each year or can use seed produced the previous year, and how much profit potential there is in producing and selling seed of a particular crop.

I think that people close to this situation are generally satisfied with the division of activity between public institutions and private firms. The division is always shifting, however, as science, technology, and economics change. The so-called "brown-bag law", which limits the sale of scientifically improved varieties by farmers to farmers, may shift the current public-private balance in the seed industry. It will probably encourage more public and private investment in plant breeding, by allowing both institutions and seed firms to realize a higher return on that investment. Those who advocate broad, sweeping policy toward either end of the spectrum of privatization views are probably not aware of the complex realities associated with these services. But how can we determine who should pay for and who should conduct these activities? It seems logical that those who benefit from food and agricultural research and extension should pay most of the costs. Identifying and measuring the real benefits and the recipients of benefits are sometimes difficult. They differ in each individual project.

WHO BENEFITS AND WHO SHOULD PAY?

Figure 1 is a conceptual framework or model that I find useful in thinking about agricultural R&D and related issues. It depicts the flow of positive benefits and negative benefits (damage) to "producers" and consumers after a useful new agricultural technology is introduced. For purposes of this discussion, I use the word "producers" in a broad sense. In a given situation, "producer" may include producers, processors, distributors, retailers, and/or firms providing inputs and support services; in other words, all those involved except the consumers of a product or service.

The first to benefit are the earliest and most effective adopters of the new technology. They realize greater profit initially because of lower costs and/or higher prices associated with an improved product or process. As soon as a significant number of producers adopt the new technology, supplies of the resulting products increase and/or prices come down and the initial benefits are competed away. Then the benefits accrue to consumers in the form of higher quality, safer, more affordable, and convenient products and services.

Studies of the diffusion of innovation (Rogers 1962) suggest that less than 15 percent of farmers fall into the category of early adopters. Other farmers adopt technology after watching the early adopters. Producers who are late to adopt the new technology may actually be placed in a less advantageous competitive position and be disadvantaged by the new technology. Eventually, the technology is replaced by newer technology and no further benefits or damage is realized from the earlier technology.

Studies indicate that over time, most of the benefits of food and agriculture research accrue to consumers (Huffman and Evenson 1993). One of these benefits is that consumers spend a decreasing proportion of their family incomes for food and other agricultural products. As a result, the nation expends less of its human and other resources to meet basic needs. Resources are freed to invest in better homes, cars, schools, universities, health care, education, infrastructure, national defense, and other non-food products and services, not to mention churches, museums, historical monuments, and other ornaments of great civilizations. Thus, food and agriculture research and education have profound national effects that go far beyond benefits generated within the sector itself.

DOES FOOD AND AGRICULTURE RESEARCH BENEFIT PRODUCERS?

Ever since the SAES, CES, and USDA-ARS were created, they have produced and transferred many new inventions and discoveries and much valuable information. Concurrently, the U.S. food and agriculture sector improved in its ability to meet basic needs for its products and services and captured major shares of global commodity markets. These changes were paralleled by a decrease in the

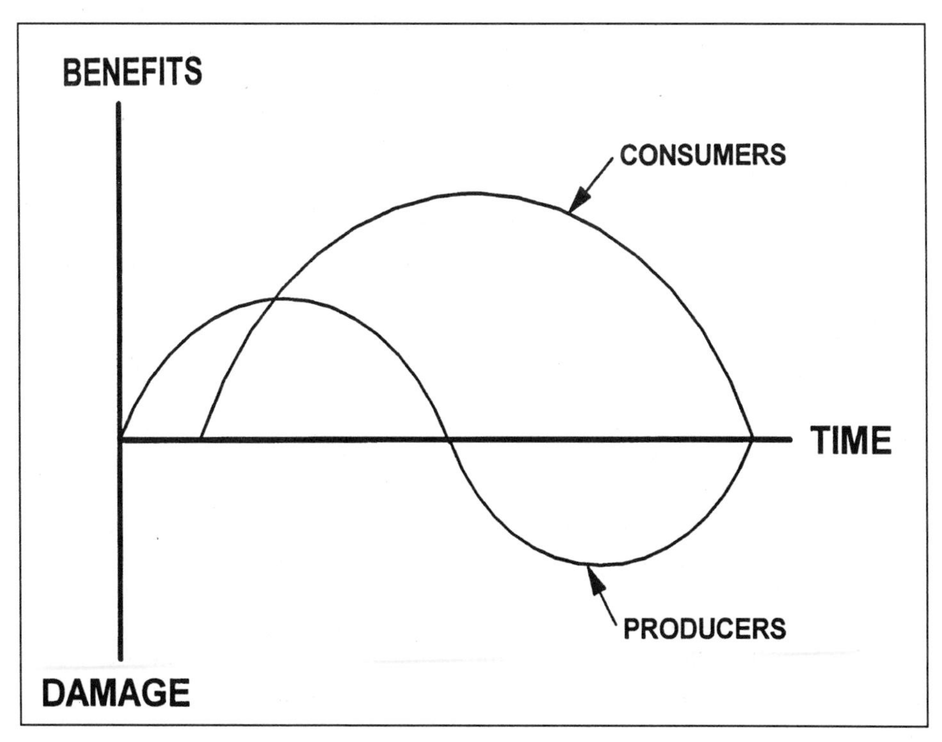

Figure 1. Graphic model of the flow of producer/consumer benefits/damage with time after the introduction of new agricultural technology.

number of farmers and, in general, a reduction in the proportion of U.S. citizens who worked in the food and agriculture sector.

It appears from the model and from history that food and agriculture R&D provide little or no net benefit for the average producer, if benefit is measured in terms of the numbers of people who are enabled to make a living in the food and agriculture sector. As is typical in market-oriented economies, when an industry matures, average profit trends toward zero, firms with less than average (less than zero) profit drop out, and the number of "producers" decreases. Food and agriculture R&D hastens this process to the benefit of consumers and the nation as a whole. In terms of overall efficiency and consumer welfare, this would seem to be a good outcome of a publicly funded activity.

There is no question, however, that certain new technologies, especially when rapidly adopted, cause dislocations that are extremely inconvenient, physically and emotionally stressful, and economically devastating for some producers who do not adopt the technologies. The public struggles with the difficult task of weighing the damage done to a few individuals against the benefits spread thinly over many individuals. A compassionate philosophy of social programs would suggest that some of the benefits accruing to early adopters and consumers should be extracted and used to help the disadvantaged competitors make adjustments.

GENERALIZATIONS CONCERNING PRIVATIZATION

A number of other hypothetical generalizations can be derived from the model portrayed graphically in Figure 1. It follows from the model and other considerations that a private firm cannot pay for R&D unless it can recover the cost of the R&D in the sale of resulting products and services. A private firm cannot recover all the value conferred on its products by its R&D, because some of that value must be passed to consumers. If consumers did not perceive that they would gain value equal to or exceeding the price, presumably they would not buy the product or service.

The early adopter benefits of any useful technology that is non-proprietary, that is, available to all producers, will soon be competed away. Because of competition, there are no net economic benefits, other than early adopter benefits, of non-proprietary technology for the producer. After a brief period when early adopters reap benefits, the benefits of specific technologies and information generated and transferred by non-proprietary university research and extension all accrue to the consumer. Thus it is appropriate that the consumer should pay for a significant portion of the programs generating these benefits. Consumers pay for public sector R&D by paying taxes.

Of course, U.S. farmers are not the only farmers in the world. If publicly financed research and education provide sustainable competitive advantage for U.S. producers competing in world markets, their early adopter benefits are extended. As individual producers, however, their principal competitors are other U.S. producers. If a state agricultural experiment station mounts strong programs of adaptive research, the results of which are specific to the unique soil, climatic, and socio-economic situations of that state, the food and agriculture sector of the state may gain competitive advantage. In this case, the information is, in effect, proprietary for the producers of that state. Whether the advantage is sustainable depends on whether the state agricultural experiment station and extension service can sustain their adaptive research programs at competitive levels relative to other states and nations, and whether they can help producers capitalize on the state's inherent advantages. SAES and CES can provide information that is proprietary because of its site and situation specificity, and because it helps farmers and other producers select among competing products and services (Holt 1987).

Traditionally, conducting adaptive research that provides competitive advantages for locales has been a major role of SAES and CES. For several decades, however, this activity has been downsized in universities in favor of basic and early-stage developmental research. The latter are regarded in some circles as more challenging, rewarding, sophisticated, and prestigious.

Major multinational private firms provide inputs for producers worldwide. In many cases, U.S. farmers do not gain competitive advantage from using those inputs, because their foreign competitors are also using them. For the same reason, farmers do not gain competitive advantage from information provided by multinational firms, except as early adopters. Theoretically, this means producers cannot afford to pay much for the information. If the early adopters have to pay for the information, it reduces their incentive to take the risks associated with being an early adopter.

This is what causes me to doubt that the private sector can or will provide all the services traditionally provided by agricultural experiment stations and CES. Most private firms provide input-related, non-proprietary information to farmers. They have to recover the cost of generating and delivering the information from the farmers who use their products and services. But because it is non-proprietary information, the farmers can't pay much for it.

Increasingly, commodity groups are trying to find ways to assure that the benefits of research they sponsor do not all spill over to producers in other states and nations, who are their competitors. Even checkoff boards sometimes are reluctant to support the routine, presumably mundane, "rate, date, and depth" experiments, variety trials, etc. (adaptive research) that are most likely to provide proprietary information for producers. Research administrators like to support more glamorous and exciting projects. However, there is probably little direct return to farmers for investments of commodity checkoff funds in research and extension that does not provide proprietary advantage for the farmers who pay the checkoff.

It is important to remember that the innovators and early adopters are not only at the leading edge. They are also at the "bleeding" edge, that is, subject to the risks of adopting new technology. Sometimes new technology is flawed in unexpected ways and may do more harm than good. If there is no good chance of reaping profit from a new approach, there will be no early adopters and thus no adopters at all. There are lots of examples of new technologies that were never adopted because nobody saw opportunities to gain competitive advantage, recover the costs of adoption, and reap a profit.

Much of the improved food and agriculture technology and information generated by land grant university research operations and disseminated through traditional extension programs is available to all producers and thus provides no proprietary advantage. Therefore, in my view, producers cannot afford to pay the costs of generating and delivering that service. Actually, university research and extension fosters competition in the food and agriculture sector, with great benefits for consumers.

The general public, as consumers, are the beneficiaries of that research and extension and, in the long run, are the only ones who can afford to pay for it. Historically, supporting SAES and CES has been an excellent use of public funds. I see no reason to expect that the return on that investment will decrease in the future.

INTELLECTUAL PROPERTY PROTECTION

Driven by the forces depicted in the graphic model (Figure 1), private firms conduct or contract for research to produce new technologies and information that are not immediately available to their competitors. They use various mechanisms, including trade secrets and patents, to protect the "intellectual property" they produce or purchase. Without some proprietary technological edge, they are vulnerable to the dog-eat-dog competition of commodity markets; that is, markets in which there are no perceived differences in value among the various sources of the product and/or service offered.

Recognizing that private firms cannot recover the costs of R&D that does not generate proprietary advantage, universities enter into agreements to license patented or otherwise protected inventions and discoveries exclusively or semi-exclusively. These licenses prevent the competitors of a firm that sponsors research from appropriating the resulting new technology as soon as it is developed. They assure sponsors that they will have a reasonable chance to recover their investment in university research and the additional private R&D required to commercialize university inventions and discoveries. Unfortunately for private firms but fortunately for consumers, proprietary arrangements may delay the time when early adopter benefits are competed away, but they ordinarily do not prevent that time from coming.

Because firms must recover the costs of research programs they sponsor, I see no reason to be confident that, if the public decided not to support food and agricultural research and education, the private sector would step into the gap and conduct the programs abandoned by the public. Many of the programs would not afford private firms competitive advantage. Evidently, there are some research and education programs that are best conducted, from an economic standpoint, by the public sector. Investment of public funds in those programs generates a high return.

EVIDENCE FROM STUDIES OF RETURNS TO RESEARCH

Huffman and Evenson (1993) summarized the results of many studies, including their own extensive work, of economic returns to public investment in agricultural research and education. They concluded that the internal, annual rate of return on the aggregate bundle of public agricultural research is 41 percent. Internal return in this case means to the investors, i.e., society or taxpayers. An annual, pretax return of 5 percent would be considered low. Anything over 25 or 30 percent would be considered excellent.

The return to agricultural research is extremely high among public sector investments, many of which yield less than 10 percent. It is difficult to separate the returns to research and extension, because of their complimentarity, but these authors estimated the return to extension for the period 1951 to 1982 to be 20 percent. They believe the return to extension decreases as farmers become better educated and are better able to get information from primary sources, including public and private scientists and educators who are not extension specialists. To the extent that extension takes the form of continuing education, one would not expect a decrease in return to investment in extension.

Rates of return are especially high (47 percent) for public research on the major commodity crops grown in the Midwest. The research includes research on utilization of these crops. The exceptionally high rates are realized because these crops are produced on such large acreages and are traded in such large quantities in domestic and international markets. A small improvement in quality, productivity, or efficiency results in great aggregate benefits.

All consumers benefit from public investment in food and agriculture research, but low income consumers are the greatest beneficiaries. This is because low-income people spend a relatively large proportion of their income on basic needs that are provided by the food and agriculture sector. Also, they pay proportionately less in taxes.

Value delivered by various entitlement programs is greatly increased and cost is reduced by food and agricultural research. People surviving on public welfare, social security, and pension benefits expend a large proportion of their limited funds to purchase basic needs, especially food. Food and agricultural research allows the poor, indigent, elderly, and retired persons to purchase more and better quality products with limited funds.

Likewise, food and agricultural research improves health and decreases the cost of health care programs. Access to high quality, nutritious, affordable food is the best insurance against health problems. This is clear from studies of the relationships among food quality, affordability, availability, and life expectancy. Other health-related effects include reduction in worker absenteeism, greater worker productivity, and better mental development in children. These are further examples of benefits of food and agricultural research that accrue to the general public.

Skeptics argue that any activity generating 41 percent annual return on investment will attract private investment and thus does not have to be supported by the public. The problem with this argument is that the 41 percent includes both producer and consumer benefits, with most of the benefit accruing to consumers. In spite of the great increase in private sector research, the returns to public sector research remain high. In most agricultural subject matter areas, the producers cannot capture enough of the 41 percent return to pay for the R&D required to generate and maintain it.

THE "MARKET" FOR AGRICULTURAL RESEARCH AND EXTENSION

According to one economic philosophy, when people advocate government intervention in a "market", they have concluded that market forces are not bringing and will not bring about the desired outcome. We can restate in these terms the argument of those who resist the privatization of public food and agriculture research and extension programs. If those programs were privatized and thus subject to typical market forces, they would not address the right issues. Instead they would address the narrow, parochial, and profit-driven interests of the private firms conducting the programs, according to this view.

Those who favor privatization, on the other hand, believe that markets discern program needs more accurately than legislators and government agencies, especially in very complex situations. They argue that if private firms cannot recover the cost of conducting agricultural research and extension programs, the programs are obviously not contributing enough value to justify the costs. If they were, consumers would recognize this value and pay for it.

But what if the producer benefits are competed away? What if the value is small for any individual consumer (taken for granted or too small for them to be aware of it) but is spread over the entire population of consumers? What if all the suppliers provide the same value in the product, so none commands a higher price? And what if that value, multiplied by the millions of people who consume improved products and services, is huge? Who should pay for the R&D that leads to that high value?

It is most useful to think about the market for agricultural research and extension as one market in which both public and private institutions, organizations, and firms are suppliers and consumers. The consumers in that market include legislators, government agencies, private firms, and individuals. Legislators appropriate and agencies allocate money for investment in public and private institutions that conduct agricultural research and extension, and in specific research and extension programs.

Likewise, private firms and individuals enter into contracts, provide gifts, or otherwise arrange with public institutions or other private firms for research and extension education programs. Or they may invest in their own internally conducted research and education programs. The competition for resources within this market strongly influences the way resources are allocated. To escape competition based only on cost, the competing suppliers try to differentiate their products and services or try to develop innovative new products and services. Perhaps the most effective strategy in this market is to form alliances. The consumers of R&D have a wide selection of options.

In each case, the legislators, agencies, firms, and individuals pay a price determined by their perception of what is fair considering the value they expect to receive. They balance their investment in those services with their perceived needs for other products and services. In these ways, the market for agricultural research and extension is like any other market.

The market for agricultural R&D, like other markets, is driven by perception, not necessarily by reality. To put this another way, the price consumers are willing to pay for agricultural R&D is not based on actual value but on perceived value. If the market does not seem to be functioning well, it may be because the customers are not well informed about the value that is or can be obtained from the service. Constituents may be more concerned, knowledgeable, and/or confidant about short-term than long-term benefits. Or perhaps they have no good way of acting on their perceptions. Or maybe the perception that the market is not functioning well is incorrect.

APPROPRIATE ROLES

Taking into account the complex relationships among public and private groups that conduct and consume agricultural R&D, the market for these services, and the model of benefits and damage, I believe the following are appropriate roles for public and private R&D programs. These are presented not as policy suggestions but as opinion based on observation. I believe the market for food and agricultural research tends to move programs in this direction.

If a research and education program helps an individual private firm or an alliance of firms achieve sustainable competitive advantage and the program costs can be recovered in the sale of products and services, clearly the firm or alliance should pay most of the program costs, regardless of who actually conducts the research. Such programs usually include firm-specific background research, product-oriented developmental research, and firm-specific technology transfer programs. It seems obvious that the costs of firm-specific market research and intelligence gathering should be borne by private firms.

One might jump to the conclusion that the costs of waste-stream management research leading to reduced environmental impact should be borne by the public, since a private firm may not be able to recover the costs associated with achieving reduced off-site impacts. However, firm-specific, waste-stream management research may lead to proprietary technology that reduces costs of operation and/or gains other efficiencies that contribute directly to the firm's profits. In this case, the private firm should pay at least part of the costs.

It is appropriate for the public to pay at least part of the cost of food and agriculture research that benefits consumers in general and for which the costs cannot be fully recovered in the prices consumers pay for agricultural products. Such programs may improve the safety and quality of products and make them more affordable. They may conserve natural resources or reduce offsite environmental impacts of agricultural operations. These programs may provide unbiased information on how competing products perform in specific sites and situations, thus enabling users of the information to select the alternatives that best fit their needs.

In general, users of information from comparison trials regard university scientists as unbiased and credible sources of information. Collectively, land grant universities have many research farms representing many different locales and therefore are in a good position to compare competing products under a wide range of soil, climate, and socio-economic conditions. This contributes to overall efficiency, because it reduces the need for each private firm to conduct its own comparisons in many different environments. Among other benefits, private firms can use university trials to see how their proprietary prototype products compare with new products from competing firms. Ordinarily they bear at least part of the costs of such studies.

Sometimes food and agriculture research and education programs can be conducted more efficiently in public institutions, because of certain economies of scale and scope enjoyed by these institutions. Sometimes the public should conduct programs because they play essential roles in the educational process. Sometimes the public should conduct or at least coordinate programs because they involve a very broad subject matter scope and a spectrum of specialists that cannot be found in a single private firm or category of firms.

Programs of developmental research that generate improved practices rather than improved products may not provide a profit opportunity for private firms and so may be more appropriate for public institutions. Programs from which private firms can capture no proprietary advantage are particularly appropriate for public institutions. Basic research, research on research methods, and public policy research are other categories that are generally conducted in public institutions, although very valuable basic research programs are conducted by some large private firms.

A question related to public vs. private funding is whether public support for food and agricultural research and extension programs should come from local, state, regional, or national sources. The concept of spillover benefits (Ruttan 1982) is applicable to this question. If individual states invest in programs the benefits of which accrue to other states or to the nation as a whole, those programs will be targets of underinvesment. Because benefits of most public agricultural research and extension projects spill over considerably from one state to others, it is justified for the federal government to invest significantly in those programs. Some benefits of U.S. food and agricultural research spill over to foreign consumers of products exported from the U.S. Likewise, U.S. consumers benefit from foreign or U.S. research that improves imported agricultural products.

A view expressed frequently, even within the public sector agricultural research establishment, is that government (the public) should support long-range, high-risk research (basic research), and the private sector should do the applied research and extension. Some leaders in recent years suggested that extension should move away from programs that directly support capital-intensive agriculture and more toward programs that address needs of what they see as a broader constituency.

These two generalizations are too broad. They are certainly not supported by the observations and the model described in this paper. Application of these generalities in political and institutional decision-making in agriculture caused some influential client groups to become disillusioned. Their withdrawal of support continues to cause problems for public institutions. Some disgruntled constituents became strong proponents of privatization and withdrew their political support for public research and extension institutions. This is how a market responds when there is a perception that less value is being delivered. A service institution should be careful not to alienate traditional clientele until new client groups are clearly identified and committed to using and paying for the services rendered.

IMPROVING MARKET FUNCTION IN ILLINOIS

Agricultural leaders in Illinois are trying to improve the function of the market for food and agriculture research programs. For the first fifteen years or so after the corn and soybean checkoff programs were established, checkoff money was invested in research and education on specific commodities, particularly corn and soybeans. In general, the relationship between the universities and the commodity organizations was good and steadily improved as they became better acquainted. Nevertheless, there was an undercurrent of dissatisfaction on the part of farmer board members.

Commodity groups solicited proposals and university scientists responded with typical competitive grants proposals outlining specific scientific objectives. The research was funded, conducted, and reported. Often the results were exciting and promising. Sometimes, the results were not immediately applicable to the practical needs and opportunities of producers. Usually, much more R&D was needed to fully implement the new information. Farmer board members wondered when they would be able to implement practical improvements that were generated by checkoff funded research. This was especially important to board members, because checkoff programs are subject to periodic referenda to determine if they will continue. Farmer boards want checkoff programs to be successful so they will continue to be funded.

This is clearly a situation in which farmers pay for university research. Their perception of value received or return on investment determines if they will invest in the programs and how much. Originally, both checkoff boards and university scientists and administrators took it for granted that good research would inevitably lead to improved technology. They assumed that improved technology would provide sustainable competitive advantage for the soybean industry in each state and in the nation as a whole. The D part of R&D was, to some extent, left to chance, which is not uncommon in university research (Smith and Tsang 1995).

More recently, checkoff boards are not leaving as much to chance. They solicit proposals in categories of desired practical outcomes. They ask applicants to state objectives in terms of specific outcomes that fall into those categories. If an applicant proposes to do basic research, the checkoff boards want to make sure someone is going to do the developmental and adaptive research and transfer the technology, so the practical goal is achieved or, at least, has a chance of being achieved. They want to make sure all the linkages between proposed research and desired outcomes are firmly established.

Checkoff boards want the programs they finance to yield sustainable competitive advantage for their producer constituents, who are supplying the money. They know that producers cannot afford to pay much for research and extension programs that do not provide proprietary access to technology. To the extent that checkoff-funded programs focus on problems and opportunities that are unique to the state or nation covered by the checkoff program, some proprietary advantage can be achieved.

Checkoff boards expect the "bugs" associated with new technology to be worked out by researchers, not left for producers. They want the results of their programs to be, in the words of a Kansas farmer, "bulletproof." They want the programs and projects to be strategically sound, fast, efficient, and thorough, so their chances of gaining early-adopter benefits are maximized.

When research and extension programs are organized around the desired, practical outcomes, sponsors are in much better position to evaluate the programs they support. They measure progress toward achieving the desired practical outcomes. This progress, rather than the number of scientific papers generated or the number of people who attended extension meetings, is the measure of success of agricultural research and extension programs. With better informed and more discriminating customers in the market for agricultural research and extension and more competition among the institutions that provide these services, the market's efficiency, that is, its ability to discover appropriate prices, is improved.

Capitalizing on experience gained in university/commodity group relationships, Illinois citizens have recently formed an organization called the Illinois Council for Food and Agriculture Research (C-FAR). The mission of C-FAR is to increase state funding for food and agriculture research and to improve the process by which constituents have input to research priority setting. The system being implemented causes C-FAR funded projects to be organized around and focused on the desired practical outcomes identified by C-FAR members.

There are other examples of situations in which universities and various groups of constituents are redesigning their relationships to improve constituents' ability to communicate their needs and expectations and universities' ability to meet those needs and expectations. The general approach is to focus on the desired practical outcomes and evaluate programs in terms of how well they achieve those outcomes. The GPRA actually mandates this approach to planning and evaluating federally financed programs.

CONCLUSION

It does not follow from this analysis that if the public stopped supporting all or some of the agricultural research and extension programs it now conducts, the private sector would pick up those programs and pay their costs. In fact, it is most likely that private firms would not be able to conduct many of these programs at a profit. It does follow that as conditions change, the balance between private and public investments in food and agriculture research and education will probably change, in both predictable and unpredictable ways.

In the future, only rarely will one individual, university, agency, or private firm have adequate knowledge, skills, resources, and relationships to conduct all the R&D required to achieve a desired practical outcome in the food and agriculture sector. A team approach will almost always be required. The teams will necessarily be interdisciplinary, cross-functional, and interinstitutional. They will almost always involve cooperation among public and private institutions, agencies, organizations, and firms.

Within almost every project, there will be roles that are clearly appropriate for public funding and public institutions and others that are clearly appropriate for private funding and private firms and organizations. As more people gain experience in planning, organizing, coordinating, executing, and evaluating outcome-focused research and education projects and programs, the function of the market for food and agriculture research and education will improve.

REFERENCES

Barry, P. J., S. T. Sonka, and K. Lajili. 1992. Vertical coordination, financial structure, and the changing theory of the firm. American Journal of Agricultural Economics, 74:1219-1225.

Chief Financial Officers Council. 1995. Implementation of the Government Performance and Results Act (GPRA): a Report on the Chief Financial Officer's Role and Other Issues Critical to the Government-wide Success of the GPRA. Chief Financial Officer Council and GPRA Implementation Committee, U.S. Office of Management and Budget.

Cooper, R. G. 1990. Stage-gate systems: A new tool for managing new products. Business Horizons, May-June, 1990.

Holt, D. A. 1994. Agricultural Experiment Stations. In Encyclopedia of Agricultural Science Charles Arntzen (ed.). Academic Press, Inc.

Holt, D. A. 1993. Changes in agriculture and agricultural institutions. p. 239-254. In U.S. Agricultural Research: Strategic Challenges and Options, R.D. Weaver (ed.). Bethesda, MD: Agricultural Research Institute.

Holt, D. A. 1991. Organizational paradigms of agricultural research and development. Proc. Fortieth Annual Meeting of the Agricultural Research Institute. Bethesda, MD: Agricultural Research Institute.

Holt, D. A. 1987. A competitive R&D strategy for agriculture. Science 237:1401-1402.

Huffman, W., and R. Evenson. 1993. Science for Agriculture. Ames, IA: Iowa State University Press.

Holt, D. A., and S. T. Sonka. 1995. Virtual agriculture: developing and transferring technology in the 21st century. In Proc. 2nd Intl. Conf. on Site-Specific Management for Agricultural Systems, P.C. Robert, R. H. Rust, and W. E. Larson (eds.). Madison, WI: American Society of Agronomy.

Jarvis, J., B. Dreben, E. Holtzman, and B. Kreiser. 1994. Corporate funding of academic research. In University-Business Partnerships, N. Bowie (ed.). Lanham, MD: Rowan and Littlefield Publishers, Inc.

Matkin, G. W. 1992. Technology Transfer and the University. New York: Macmillan.

Pray, C. 1993. Trends in Food and Agricultural R&D: Signs of Declining Competitiveness? In U.S. Agricultural Research: Strategic Challenges and Options, R.D. Weaver (ed). Bethesda, MD: Agricultural Research Institute.

Rogers, E. M. 1962. Diffusion of Innovations. New York: Free Press of Glencoe.

Ruttan, V. W. 1982. Agricultural Research Policy. Minneapolis, MN: University of Minnesota Press.

Smith, T. P., and J. C. Tsang. 1995. Graduate education for research and economic growth. Science 270:48-49.

Section Two

ORGANIZATIONAL AND TECHNOLOGICAL CHANGE IN PRODUCTION SYSTEMS

5

Can Cooperatives Survive the Privatization of Biotechnology in U.S. Agriculture?

Elizabeth Ransom, Lawrence Busch, and Gerad Middendorf

Most agricultural cooperatives emerged out of disputes between farmers and input suppliers and processors. Farmers banded together to obtain needed inputs and to sell farm products at a fair price. Using the buying and selling power of the cooperative, farmers were able to combat what they saw as predatory pricing on the part of local monopolies. Indeed, it was in response to this concern that Congress exempted cooperatives from antitrust laws under the Capper-Volstead Act of 1922 (Young 1991). They could band together to buy inputs in bulk directly from manufacturers, sell their products directly to processors, and even limit production through marketing orders, thereby increasing prices. In some cases, cooperatives even entered the input production and output processing businesses, providing for farmers the kind of vertical integration that had proven so successful to their non-farm competitors.[1]

Throughout their existence, agricultural cooperatives and private input supply firms have traditionally relied on research and information from public research institutions, i.e., land grant universities and USDA laboratories. Although nearly all mechanical and much chemical research had long been the province of the private sector, most such research focused on generic products that are resold through both cooperative and investor-owned suppliers. Public biological research

The research reported in this paper was supported by a cooperative agreement between Cooperative Services, Rural Development Administration, USDA and the Michigan Agricultural Experiment Station. The opinions expressed herein are those of the authors.

ISBN 1-57444-104-3/98/$0.00/$.50

also focused on undifferentiated products that were made available to all comers including cooperatives. Those cooperatives that have engaged in processing have tended to limit their activities to those products requiring little post-harvest handling (e.g., fresh produce) or for which processing technologies are well-established (e.g., dairy). High value-added products for consumers have, in large part, been left to the private sector. Large food companies have long supported research and development programs, but they have tended to focus on product differentiation rather than scientific research.

More recently, however, private firms have made substantial inroads into all forms biological research in agriculture (Kloppenburg 1988). The development of the new biotechnologies in agriculture, unlike previous research-intensive initiatives, is largely concentrated in private industry, with 71 percent of all field trials currently being conducted by the private sector (Ahl Goy and Duesing 1995). This is due in part to significant changes in the legal framework for research and development (Busch et al. 1995). In particular, the passage of the Plant Variety Protection Act in 1970 gave seed companies greater control over the results of research efforts and spurred the purchase of those companies by chemical and pharmaceutical giants. Later, in the *Diamond vs. Chakrabarty* (447 US 303) case, the U.S. Supreme Court extended utility patent laws to cover lifeforms. In 1985, as a result of *ex parte Hibberd* (227 USPQ (BNA) 443), the Patent Office began to grant utility patents for plants. In 1991 the international convention governing plant variety protection was revised, broadening the definition of a variety and thereby making protection claims broader. This strengthening of intellectual property rights has created an incentive for those few large private firms that are engaged in research to follow their products further downstream—sometimes as far as retailing the final product—so as to maximize the value added for the company. In practice this means more direct marketing of inputs (e.g., Monsanto's direct marketing of bovine somatotropin (BST) to farmers) and contracting directly with producers.

At the same time, the granting of federal tax deductions for venture capital investments in research under the Economic Recovery Tax Act of 1984 spurred private investment in and control over biological research, while the extension of patent rights to universities operating with federal funds under the Federal Technology Transfer Act of 1986 has made universities arguably less likely to release results to all interested parties. Furthermore, many university and USDA researchers have begun to see input suppliers and processors as their major clients, to the detriment of farmers, cooperatives, and small farm-related businesses (Busch 1990; Weaver 1993).

In the 1980s when the "new" biotechnologies were still heavily concentrated in research and development labs, there was much speculation about the effects biotechnology would have on agricultural cooperatives (Lacy and Busch 1988;

1989). Questions were raised such as: What biotechnologies will be developed and for what reasons? Who will control access to the product once it is developed? Will the decline of publicly-supported research increase the pressures on cooperatives to form alliances with or even transform themselves into investor owned firms (IOFs) so as to better compete in the marketplace?

Today, just as in previous years, opinions regarding the effects biotechnology will have on agricultural cooperatives vary according to where one is located in the agricultural system. One unifying factor of most discussions concerning cooperatives and biotechnology is the need for cooperatives to play active roles in shaping the future of biotechnology in agriculture. In order for cooperatives to play active roles in shaping their own future, the dominant response among economists advising cooperatives has been to recommend the development of strategic alliances with investor-owned firms or, more rarely, with other cooperatives (Torgerson 1993). As they struggle to adapt to a changing environment, many cooperatives have been forced to redefine their goals or—at the other extreme—to totally abandon the cooperative approach and privatize their operations (Collins 1991; Reilly 1991; Schrader 1989a, 1990).

In this paper we begin by discussing the current state of cooperatives. This is followed by a discussion of the emerging biotechnologies that are being developed for plants, food, and animals. Finally, we discuss the import of alternative strategies for the future of agricultural cooperatives.

THE STATE OF COOPERATIVES TODAY[2]

There are three classes of agricultural cooperatives: marketing, supply, and service. Marketing cooperatives is the largest category, with 2,218 classified by the Agricultural Cooperative Service (ACS) in 1992. Supply cooperatives is the second largest category with 1,618 listed in 1992. Service is the smallest category, but the only category which has seen an increase in numbers in recent years. This is due primarily to reclassification of some cooperatives formerly considered to be engaged in marketing. There were 479 service cooperatives in 1992.

Overall, the total number and the total membership in cooperatives has been on the decline over the past decade. There has been an average decrease of about 200 cooperatives per year. This reflects the overall decline in the number of farms in the U.S. (see Figure 1). Cooperatives numbered 5,989 in 1983, but as of 1992, the number had fallen to 4,315. Membership also fell from 5 million in 1983 to 4.1 million in 1992. Consider why these changes have occurred.

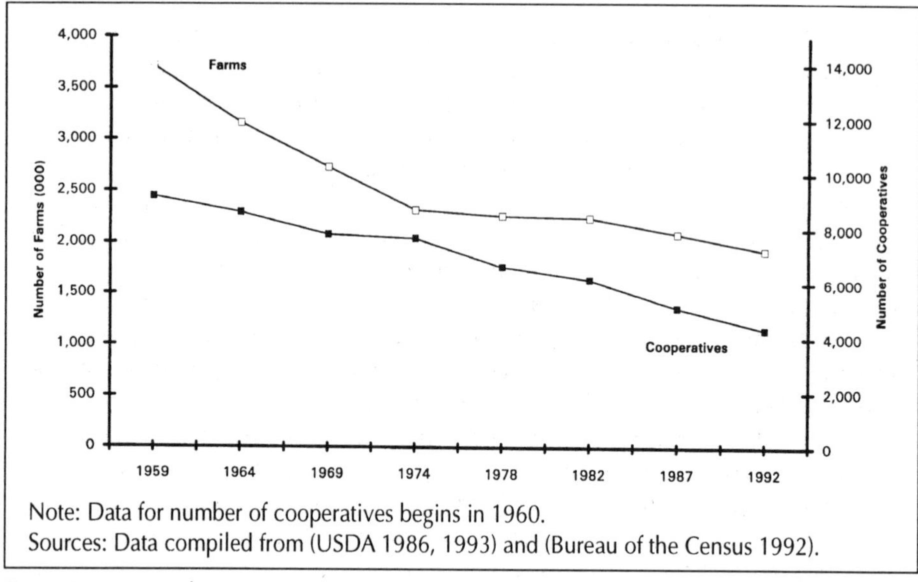

Note: Data for number of cooperatives begins in 1960.
Sources: Data compiled from (USDA 1986, 1993) and (Bureau of the Census 1992).

Figure 1. Farms and Cooperatives, 1959-1992

First, many individual cooperatives, which in the past served a local community, have been acquired, merged, or consolidated with other cooperatives for financial or other reasons. These cooperatives have then been operated as branches, which provide service to members who are further away from the cooperatives' headquarters. The significant factor is that the larger cooperatives operate many more branches than the smaller ones. For example, 0.5 percent of the grain and farm supply cooperatives accounted for 33.5 percent of the total number of branches operated by such cooperatives.

Second, most cooperatives continue to be small and serve local areas, but there are several that are increasing in size and expanding across regional, national and international markets. In 1992, the 100 largest agricultural cooperatives accounted for 54.9 percent of the total business volume generated by cooperatives. Moreover, there are two types of small cooperatives: traditional cooperatives (e.g., small dairy cooperatives) as well as new cooperatives with closed memberships that are designed to capture small market niches ignored by both large coops and large IOFs (Torgerson 1994; Kibbe 1995).[3] This parallels the general trends in concentration and specialization in agriculture.

Third, the role and function of cooperatives is dependent upon the nature of farming in each region. For example, bulk seed sales are greatest in the Midwest, where large farms have large acreages devoted to corn, soybeans, and wheat. Not surprisingly, seed companies are very active there. Under these circumstances, "if the seed company is willing to sell directly or indirectly to the farmer, there will be little incentive for farmers to purchase seed from their cooperative or other farm supply stores" (Chesnick and Staiert 1992). In contrast, in the Northeast, Southeast, and South the types of crops and the size of acreage planted vary widely and in turn, "most seed sales are accomplished through cooperatives and farm supply stores" (Chesnick and Staiert 1992). Cooperatives may be more successful investing in biotechnology for seed production in these areas than in the large farm agriculture of the Midwest where corporate players are likely to dominate the scene.

Fourth, many of the larger cooperatives have begun to resemble IOFs. Indeed, as Gray (1995) has noted, much of the literature on the economics of cooperatives treats them as a special case of IOF. As he correctly notes, this assumes that cooperatives are organized around the sole goal of maximizing or optimizing their rate of return to investors. However, farmers create and join cooperatives for reasons other than profit maximization by the cooperative. Issues of market power (gained by collective action as a cooperative), local control over land and labor (cooperatives are place-bound in ways that many IOFs are not) and democratic decision making (one member, one vote) are also of concern to members.

Many cooperatives were formed because farmers were unable to get the services they needed at prices that they considered fair. By creating a cooperative, they were able to extend their locus of control beyond the farm gate while still remaining farmers. The history of cooperatives shows that farmers often found themselves at the mercy of a single purchaser of farm products or supplier of inputs that had a local monopoly. Farmers were able to use cooperatives to permit economies of scale in purchasing that were far beyond the resources of individual members. Cooperatives enabled farmers to maintain at least some of their traditional independence in the face of large, corporate entities that sought control over input and output markets.

Thus, cooperatives are hybrid forms. They are neither traditional IOFs nor are they non-profit organizations. Their members are simultaneously their customers *and* their investors. Unlike IOFs where the size of one's investment often determines the voting power, cooperatives operate on the basis of one member, one vote. Also, cooperatives pay patronage refunds in proportion to the business conducted with the cooperative.

From the perspective of most members, cooperatives are always secondary and linked businesses. They are secondary in that most farmers see cooperatives as ways of protecting their primary businesses—their farms—against fickle suppliers or marketing and processing companies. For example, dairy farmers see their cooperatives as reliable places to sell their milk and to purchase farm inputs. They are reliable in that they will always serve all members and they will not pull out of an area even if total profits might be higher elsewhere. They are linked in that the success of the farmer-members at farming and the success of the cooperative are not fully separable. Cooperatives form part of the same commodity subsectors, or provide essential inputs for the commodity subsectors in which their members participate. They can be viewed as a form of loose vertical integration pursued collectively by farmers.

As farmer-members, farmers wish to optimize farm and cooperative profits simultaneously. In doing so they may be more than willing to sacrifice cooperative profits in order to preserve or enhance profits on the farm—if that will optimize their total profits. Moreover, since not all farmers are in the same financial condition, they will likely argue over the proper balance of farm and cooperative profits.

In order to maintain this balance, cooperatives and their farmer-members have traditionally relied on public research from land grant universities as noted above. Indeed, it is because of public research that many cooperatives were created. For example, Cropp and Ingalsbe (1989) note that:

> "...cooperatives have been the predominant organizational approach to improving beef and dairy cattle through artificial insemination. After experimentation, largely at land grant universities, in the late 1930s proved that the practice held great potential, many [cooperative] breeding associations were formed."

The current decline in publicly supported research and extension and the concomitant growth of private research—especially biotechnological research—may well be detrimental to farmers and their cooperatives. With this in mind, let us consider some of the effects the new technologies may have on cooperatives in more detail.

THE NEW BIOTECHNOLOGIES

There are three major categories of new agricultural biotechnologies: plant, food, and animal. All are distinguished by the fact that they permit far greater precision and speed than older technologies. This, combined with the broadened intellectual property rights described above, permits market differentiation and segmentation of a kind previously found only in certain branches of manufacturing (Urban 1991).

Many biotechnologies are now reaching the production stage. Of particular importance are transgenic plants (Busch et al. 1991). These have one or more features of interest to farmers and/or processors. Qualities which are being developed include pest resistance, herbicide tolerance, shelf life, and fatty acid composition. Herbicide tolerant crops, such as, corn, tobacco, and soybean, were the first group of new agricultural biotechnology products to become available to farmers. With an estimated annual value of $2.1 billion by the year 2000, 30 to 50 percent of industry research and development spending on biotechnologies is concentrated on herbicide-resistant crops (HRCs) (Krimsky and Wrubel 1996; Lee 1993.)

A recently released pest resistant crop is Monsanto's Bt (*Bacillus thuringiensis*) potato, Newleaf ™ (NatureMark 1995). Bt, which first became available as an insecticide in the United States in 1958, is a soil bacterium which creates a crystalline particle during spore formation that contains proteins called delta-endotoxins. Forty Bt crystal protein genes have been sequenced, with many of the proteins having highly specific insecticidal activity (Krimsky and Wrubel 1996). Bt appears to be effective only in insects and is harmless to mammals. Through the use of a particular Bt crystal protein, NewLeaf™ is resistant to the Colorado potato beetle. Also, recently approved by the EPA are Ciba-Geigy's Bt corn and Monsanto's Bt cotton, both of which are expected to be available commercially by the end of 1996. How these products will be marketed to farmers is as yet unclear.

A final type of plant biotechnology, and the one likely to impact cooperatives the most, is identity preserved crops (IPCs). IPCs have been engineered with specific altered traits. In order for the developers to maximize profits, they must maintain ownership of the product from the seed to the final product. In most cases this will take the form of the developer contracting out to farmers. One of the first IPCs to be marketed was Calgene's Flavr-Savr™ tomato. Calgene also recently commercialized Laurical™, a high lauric acid canola. The Flavr-Savr™ will increase shelf-life of fresh tomatoes and will also be used in tomato paste and sauces, while Laurical™ will be used in the oil, soap, and cosmetics industries.

Food biotechnologies include improved means of fermenting foods, new forms of food ingredients, flavors and fragrances, and systems that allow rapid detection of spoiled foods (Cheetham 1993; Law and Mulholland 1991). The first genetically engineered enzyme to be used in fermented foods is chymosin, a recombinant form of rennet, which is an enzyme found in the fourth stomach of an unweaned calf. Rennet is essential in cheese making as a coagulant. It was approved for use by the FDA in 1990. Research within the last few years also suggests new alternatives for the expression and production of protein engineered variants of chymosin. Future biotechnological techniques in cheese making which are expected to be of importance to the industry include recombinant starter cultures to improve flavor production and anti-pathogen activity, and protein engineering to modify the activities of enzymes which have coagulating and ripening functions (Law and Mulholland 1991). Cooperatives with dairy processing facilities will doubtless want access to these patented new technologies.

Animal biotechnologies include various growth hormones, and transgenic animal technologies that might be used to produce leaner pork or goats with pharmaceuticals in their milk. Thus far, bovine somatotropin (BST) is the only animal biotechnology which has been commercially released. BST is a polypeptide hormone which is synthesized by the anterior pituitary gland in cows. The major physiological function of BST is to increase milk production by directing nutrients away from storage in body tissue. In the early 1980's, the ability to microbially reproduce BST through recombinant DNA techniques was developed. BST injections maintain a cow's milk production at or near the lactation peak, increasing overall milk production. The amount of increased milk production varies with the type of cow, the lactation period of the particular cow and farm management. BST is currently administered to cows by repeated subcutaneous injections during lactation. Cooperative members often use BST, but to our knowledge no cooperatives are distributing it.

Porcine Somatotropin (PST) can now be produced through recombinant DNA techniques. The major physiological function of PST is to increase the average daily weight gain of pigs and to create higher quality carcasses by decreasing back fat. Compared with BST, very little debate has surrounded PST. Some reasons for this are:

1. the safety of the final product for human consumption has been confirmed;
2. hogs are usually part of a more diversified production system (although this is changing), unlike dairying, which usually occupies a more central role in farm operations;
3. hog production and pricing are not tightly controlled by federal programs, as is dairying; and finally,
4. PST has not yet been approved for commercialization because of health complications in pigs administered PST. If it is successfully commercialized, cooperatives may be bypassed.

The area which would have the largest impact on the agricultural sector is the creation of genetically modified animals for food and pharmaceutical purposes. Current developments in transgenic animals include: pigs with the human growth hormone gene; livestock designed to tolerate extreme climatic conditions; transgenic sheep that grow faster than normal sheep; sheep and cow's milk which have modified food or pharmaceutical properties; chickens with growth genes of cows; and fish that grow faster, grow bigger, and can survive better in different environments (Krimsky and Wrubel 1996; Lee 1993). There is also research being done to make animals disease resistant. This would be particularly beneficial for farm animals, which are highly susceptible to the rapid spread of diseases due to close contact with other animals (Lee 1993).

Currently, these animal biotechnologies, with the exception of BST, are far from the production stage. Although many developments and experiments are occurring in the lab, it is not clear when other animal biotechnologies will be released on the market. It is likely that the first products available commercially will be in pharmaceutical applications.

Although the new biotechnologies cover a broad spectrum of research and each product group provides its own unique set of issues, there are also some issues that are relevant to all the new biotechnologies. The issue of health and safety (to people, to animals, and to the environment) is one the greatest concerns. As industries develop these technologies, many fear that the best interest of the public will be sacrificed in the pursuit of profit. Accidental release of organisms that may create new pests, overlooked long term health problems, and the quality of life of farm animals are among the many concerns relating to health and safety.

Another area of concern is the meanings humans give to food. Rather than just viewing food for simply its nutritional value, we attach religious, sacred, and social meanings to food. A recent example was the public response to the use of BST (Busch 1991). Many view milk as one of the last "pure" foods; they see BST as adulterating the milk. While it appears that the BST controversy is over, or at least dormant, it underlines the need for cooperatives to be aware of the issues associated with consumer acceptance of modified food products. Investing in products which provide great advantages to farmers and/or cooperatives, may not necessarily be perceived in the same manner by the general public.

Yet another set of issues surrounds regulation. Regulation for safety as well as for ownership is still in the early stages of development at the national level, while the whole realm of international regulation remains virtually untouched. Safety regulation may impose high costs for cooperatives that wish to enter biotechnology product markets. Changing intellectual property rights may also increase costs for cooperatives. Indeed, smaller IOFs as well as cooperatives may find themselves excluded from participation in most biotechnology research and development.

Due to the wide spectrum of technologies being developed and the variety that exists among cooperatives, it is difficult to summarize the effects these technologies will have on cooperatives. Nevertheless, some generalizations are possible. As Figure 2 makes clear, agricultural biotechnologies provide both potentials and constraints for cooperatives. All of these issues, plus others, pose difficult deci-

Biotechnologies	Potentials	Constraints
Plant	• Serve as intermediaries between farmers and industry.	• Only co-ops with large budgets can be involved in product development, due to high cost of research. • May be locked out of markets due to contracting of specialty crops directly with farmers (e.g., Flavr Savr™ tomatoes).
Food	• Profit from the use of some food technologies. For example, the use of artificial rennet (chymosin) has benefitted cooperative cheese production.	• Decrease the share of value added received by farmers for farm products. • The development of substitutes may decrease the value of farm products (e.g., reduction of sugar prices with development of high fructose corn syrup).
Animal	• Serve as the intermediaries for products which will enhance animal production. • Open new markets for farmers (e.g., "pharmafoods"). • Can play a minimal role without detrimental effects (e.g., BST).	• Public controversy may force cooperatives to take a position on the use of a product. • Transgenic animals may be grown under direct contracts with producers, bypassing cooperatives.

Figure 2. Potentials and Constraints for Cooperatives of Private Control of the New Biotechnologies

sions for agricultural cooperatives. The ability of cooperatives to develop institutional innovations and strategies will be key. Of course, the strategies which will best serve cooperatives will vary greatly. Let us consider some of those strategies.

STRATEGIES FOR COOPERATIVES

What should cooperatives do in the face of the new private sector role in biotechnology? What strategies are likely to best serve their members? Which areas are cause for concern? How should cooperatives respond? The biggest area of concern has been the rise of contract production. Just as contracting has taken on a substantial role in chicken and hog production, there is increasing use of contracting in identity preserved specialty crops, and value-added biotechnology products. In a 1992 Indiana survey, "15 percent of farmers said they were growing on contract, and it is predicted that this will double by the year 2000" (Johnson 1995). Contracting is now rapidly growing within the seed industry. For example, Calgene's Flavr-Savr™ tomato and herbicide resistant cotton seeds, and Monsanto's Bt potatoes are grown under contract. As seeds are developed with altered characteristics and as higher and more uniform standards are demanded for raw materials for value-added products, companies are increasingly contracting out to growers. One economist notes that if manufacturers choose to establish their own distribution system, "the likely result will be the same as with hybrid seed corn. Since, the introduction of hybrids, cooperatives' market share and net margin have steadily eroded" (Ratchford 1990). Of course, in principle, cooperatives could become bargaining agents, negotiating contracts with agribusiness companies for their members. Whether they will be able to do so remains to be seen. However, their effectiveness will be in part a function of whether other similar climatic zones are available for production. If they are, then agribusiness firms might simply move production to areas where farmers are unorganized, much as manufacturing has moved in recent years to areas within and outside the U.S. where labor is unorganized. Furthermore, the larger cooperatives appear to have the upper hand. They have far more operating branches, and thus they provide industry with access to more farmers. Alternatively, larger co-ops might attempt to compete directly by developing their own specialty seeds, either independently, with other cooperatives, or in conjunction with a large seed company, but in doing so they may be pushed even further down the road to becoming investor-owned firms.

Currently, cooperatives are in a strategically important position, in that they are a leading production information source for farmers (Coffey 1993). Many cooperatives have chosen, or may be forced to choose, to align themselves as the intermediary that facilitates the connection between the processor and the producer (Ward 1993). Cooperatives are viewed as natural links between farmers and input suppliers and processors; thus they can also be a link for educational purposes. Growmark, a seed cooperative, has chosen to fill this role. Growmark states, "we have decided that to develop the technology ourselves would require too great an investment . . . . Instead, we believe our strategic advantage is to market biotech discoveries" (Barwick 1992).

Part of Growmark's strategy is to position itself as a marketing specialist within the industry (Chesnick and Staiert 1992). Other cooperatives may well follow its lead. Cooperatives can become a comprehensive distribution system, create a "one-stop solution center," and maintain their reputation as information sources. Essentially, as one article states, "cooperatives are evolving from being an off-farm arm to being an on-farm hand" (Coffey 1993). If cooperatives do not aggressively pursue this role, then "it seems likely that as agribusiness firms come to play a more central role in the development of biotechnology innovation, they will also emerge as the primary sources of consumer information about these products" (Hoiberg and Bultena 1990). Indeed, agrichemical manufacturers and other agribusiness firms are now positioning themselves to take on that role—even to the point of replacing some of the functions currently filled by the Cooperative Extension Service. However, cooperatives could play an important role for independent farmers by using collective bargaining to get the best contract for the farmers in a cooperative (Chesnick and Staiert 1992; Coffey 1993; Kidd and Dvorak 1995).

Moreover, although cooperatives may actively pursue the role of distributor, this is still largely a reactive stance to the emerging biotechnologies. What about those cooperatives that choose to compete in the development of biotechnology products? There seems to be a general consensus that cooperatives in their traditional forms cannot compete in the development of biotechnologies. The capital requirements associated with bringing biotechnological research to the final stages of product development indicate that most cooperatives will not be primary developers (Boutwell 1994).

The strategies suggested to prevent being left out of the development process have varied. One approach has been to form alliances with other cooperatives. In 1987, Cenex and Land O'Lakes began a joint venture in agronomy operations wherein fertilizers and chemicals are supplied jointly to each cooperative's members (Torgerson 1993). By expanding and strengthening their resource base, the Cenex/Land O'Lakes alliance has been successful in furthering the potential of the two cooperatives in research and development of biotechnology. While the discussion of cooperatives merging or forming alliances is not new, the new biotechnologies have reinvigorated the idea. The main reason for doing so is to pool resources so as to better compete with other large actors in the marketplace. For example, "if cooperatives were to combine their seed operations, they would have the world's fifth ranking seed company" (Coffey 1993). Moreover, cooperatives could pool their resources together to create biotechnology research and development laboratories as several have already done. It has also been suggested that cooperatives renew their traditional role of supporting public research institutions. Through marketing agreements, cooperatives might obtain exclusive rights to research results from public institutions (Chesnick and Staiert 1992).

Another strategy which has been pursued by cooperatives is joint ventures with private businesses. For many cooperative leaders and experts this seems to be a solution for survival in the future. By aligning themselves with businesses, cooperatives may gain the needed capital to pursue research. They may gain privileged access to the products developed by large corporate actors. They may also gain access to bigger markets. Examples would be West Central Cooperative's joint venture in soybeans with the Nichii Company, a $9 billion corporation in Japan (Seaman 1993). Similarly, the Sunkist Cooperative has a license/partnership agreement with Haitai Beverages, a large food and beverage company in Korea. Sales of Sunkist-licensed products by Haitai have grown from $2 million in 1986 to $256 million in 1991 (Briggs 1992).

In addition to consolidations, mergers, and joint ventures, there has been an increased use of subsidiaries by cooperatives. A subsidiary is "any cooperative or non-cooperative corporation controlled by another cooperative or non-cooperative corporation" (Reilly 1991). Reilly explains that a subsidiary can either be operated as a cooperative or as an investor-owned firm. There are many reasons for creating a subsidiary. For example, business and legal obligations are separate from the parent cooperative; therefore losses, liabilities, and lawsuits do not directly effect the parent cooperative, while profits can be transferred to the cooperative. Furthermore, subsidiaries can assist in providing functions which are quite different from those provided by the parent cooperative. In addition, subsidiaries offer the ability to use specialized management without affecting the original, parent cooperative management structure. Among the cooperatives that use subsidiaries are Welch's, Gold Kist, Farmland, and Land O'Lakes (Collins 1990; Reilly 1991).

Reilly notes that all of the strategic alliances employed by agricultural cooperatives (consolidation, mergers, joint ventures, and subsidiaries) are normally associated with investor-owned firms. He notes that, "cooperatives have used these organizational changes . . . [and] in the process, they have become larger, more complex, and more dependent on sophisticated, specialized management" (Reilly 1991).

As cooperatives adapt to a changing environment and many form strategic alliances, there is much concern that they may lose their distinctive qualities. For example, many see the business structure of cooperatives as having shifted to a structure which resembles IOFs, with more emphasis on profit and less on member services (Campbell 1993). In fact, many cooperatives, or parts of cooperative operations, have restructured into IOFs (Collins 1991; Shrader 1989a). Often cooperatives choose to transform themselves into IOFs because of the opportunity for members to obtain equity capital. The cooperatives which are primary candidates for restructuring into IOFs are those that are valued at a level above their book value. In these cases, by restructuring members may receive large payments in return for their shares in the cooperatives' equity. What should not be forgotten, however, are the benefits which member-patrons receive from a cooperative that are not defined within the framework of equity.

In a recent study of cooperatives which have restructured as IOFs, Schrader (1989a) observed that the initiative for restructuring originated either from management or from an outside expression of interest. Only after members were presented with an offer which reflected the market value of the cooperative, did the members vote to sell. Schrader emphasizes the need for cooperative management to explore alternatives *before* a cooperative is faced with a bid, and then to present members with the full range of alternatives before a decision is made.

There is also the concern that as more cooperatives opt to align themselves with businesses, instead of other cooperatives, the ability of non-business-aligned cooperatives to function and compete in the market will be sacrificed. As one USDA administrator states, "first priority ought to be given to coordination with other cooperatives because it is through this strategy that producers' position in the marketplace is united and strengthened" (Torgerson 1993).

As Kidd and Dvorak (1995) state, "by not being proactive, those in the . . . agbiotech industries are permitting activism, negative press, and just plain bad luck to sour these innovations in the minds of the public." This is true for most cooperatives as well. Their leadership has often taken a neutral stance or a wait and see approach to the products of the new biotechnologies. This is hardly surprising given the general fiscal conservatism of their farmer-members. As a result of this conservatism, as well as the high risks involved and the lack of confidence and negative views among segments of the public, many cooperatives and their leaders have opted to take a more reactive role. For example, in reference to BST one dairy cooperative leader stated, "It is not that we did not think the product was safe; we knew it was safe. It was the controversy in the public opinion that kept us from immediately endorsing it" (Personal communication 1995).

Availability of information is another issue when discussing biotechnology development. Unlike the public sector research of the past, most biotechnology research and development has been conducted in secrecy within private industry, and discoveries are protected by patents. Schrader (1989b) points out that areas which require a high level of secrecy may be problematic for cooperatives. "Cooperatives will have some problems because their affairs tend to be more public than those of other firms, and the process of informing members means revealing information that may be useful to competing firms" (Schrader 1989b).

An issue of importance to cooperative strategies is the effect of their size and power. The largest cooperatives have been the most likely to restructure. It appears that the size and strength of a cooperative will play a role in determining how effectively it can cope with biotechnology. It has already been established that in order for cooperatives to be proactive in the development of biotechnologies they must align with others to gain the needed resources to do research. It seems likely that the larger cooperatives would have more opportunities to create or enter into alliances. However, this does not mean smaller cooperatives do not have the potential to be meaningful participants.

Some cooperatives, especially smaller ones, may find that developing market niches outside the mainstream might be the most viable of strategies. In particular, given public concern over some aspects of biotechnology and the fascination with "natural" and "organic" foods and direct marketing among certain segments of the public, some cooperatives might be able to compete better by avoiding biotechnology and its products entirely. Clearly, students of the cooperative movement should examine the implications and consequences of this approach.

The push for cooperatives to become bigger to be successful in the development and use of these new technologies brings both advantages and disadvantages to farmers. Among the advantages are that farmers maintain an organization to grow with and to negotiate with private industry. If a cooperative maximizes its capabilities, it could provide farmers with an opportunity to have a choice in dealing with private industry, such as being able to choose from more than one contract and deciding which contract is the best offer. Thus, the cooperative could provide protection for farmers against local monopolies.

In the case of cooperatives with successful research and marketing strategies, farmers benefit with easy access to markets and confidence that they will receive a fair price. Yet, with cooperatives pushing toward ever larger size, the disadvantages to farmers will most likely be declines in member service and other qualities which distinguish cooperatives from IOFs. Indeed, there are many farmers who would argue that the larger cooperatives no longer are true to their cooperative ideals. Even scholars who view cooperatives from a strictly financial perspective raise the question, "Can a cooperative be too successful to survive as a cooperative?" (Schrader 1990). Unless countervailing structures are developed, as cooperatives increase in size and operate more and more like IOFs, medium and smaller size farms may be pushed to the side, while the larger farms reap most of the benefits.

As cooperatives become larger and require more professional managerial capacity, they often drift away from their farmer members. Thus, farmers lose their ability to exert influence over both the research process (as it is shifted more and more to the for-profit sector) and over the very cooperatives that are supposed to represent their interests. Moreover, with changes in the structure of agriculture, client groups that have the power to influence the direction of research are also changing (Lacy and Busch 1988). Research problems which may be of interest to a cooperative and its farmer members, may not be very high on the corporate research agenda. Many would cite BST as a recent example of a technology not developed with the best interest of dairy cooperatives or dairy farmers in mind, but rather developed by and for private industry's interest.

Since cooperatives are secondary businesses linked to farm operations, they are place-based organizations. While overt awareness and concern for the land has not always been apparent within cooperative operations, there has always been an underlying, almost taken for granted tie to the land. In contrast, the large IOFs that dominate the input supply and food processing industries are not tied to a particular group of farms or farmers and therefore they are not place-based. Unlike cooperatives, these IOFs have the ability to move to a more profitable region, a locale more environmentally forgiving, or even a location with weaker environmental laws.

CONCLUSION

"By the 21st century, 90 percent of commercial agricultural output [in the U.S.] will be concentrated in the hands of fewer than 50,000 farm firms, franchises, coops, and holding companies" (Cleberg 1995). As the number of cooperatives continues to decline, those cooperatives that continue to survive will be the most willing to change or adapt to the shifting environment. As a former CEO of Land O'Lakes states, "The mission has not changed, the environment has changed." The mission is still to satisfy the needs of the members of the cooperative (Hofstad 1990). What cooperatives may need to do to prosper in a competitive environment (if that is the goal) may or may not be in the best interests of the farmers they ostensibly represent. As Schrader (1989b) writes, "A cooperative's election to serve only the traditional medium-to-small farmer would indicate acceptance of a diminishing role."

Finally, it should be noted that generalizations about cooperatives should be avoided. For example, the release of BST, many have looked towards dairy cooperatives in an attempt to make broad statements about the effects of biotechnology on all cooperatives. However, as Ratchford (1990:35) explains, dairy cooperatives are unique in certain respects: "Together [they] have more than a 70 percent market share; the industry is vertically integrated upward by farmers; it is domestically oriented..." These qualities highlight the need for development strategies aimed at specific types of cooperatives with the effects of specific types of biotechnology in mind.

Nevertheless, all cooperatives must confront privatization and decline of public support for research. These developments will put pressure on all cooperatives to transform their organizational structures. Whether they are capable of remaining competitive with IOFs without fully abandoning the ideals of the cooperative movement remains to be seen.

NOTES

[1] Marion (1986) notes that cooperatives have been more successful in upstream than in downstream integration. This may be due in part to the (until now) relatively undifferentiated character of inputs and the important role that product differentiation plays at the retail level. The new biotechnologies are likely to increase differentiation at all levels of the agrifood system, making cooperative integration more complex and more difficult than it was in the past.

[2] All information in this section comes from USDA (1992) unless otherwise noted

[3] Unlike traditional cooperatives, these new cooperatives require a considerable initial capital investment from each member.

REFERENCES

Ahl G., P. Duesing, and J. H. Duesing. 1995. From Pots to plots: Genetically modified plants on trial. Bio/Technology 13: 454-458.

Barwick, S. 1992. Biotechnology: What we are doing at GROWMARK. American Cooperation 1992. Washington D.C.: National Council of Farmer Cooperatives, 217-18.

Boutwell, W.A. 1992. The challenges and opportunities facing cooperatives. The Cooperator. Aug. 6-7.

Briggs, G. 1992. Cooperative franchising and licensing. In American Cooperation 1992. Washington D.C.: National Council of Farmer Cooperatives.

Bureau of the Census. 1992. Census of Agriculture. Washington D.C: U.S. Dep. Commerce.

Busch, L. 1990. The social impact of biotechnology on farming and food production. In Agricultural Biotechnology, Food Safety and Nutritional Quality for the Consumer, J. F. MacDonald (ed.). Ithaca, NY: National Agricultural Biotechnology Council, Report 2.

Busch, L. 1991. Risk, values, and food biotechnology. Food Technology 45: 96, 98, 100-101.

Busch, L., W.B. Lacy, J. Burkhardt, and L. R. Lacy. 1991. Plants, Power, and Profit: Social, Economic, and Ethical Consequences of the New Biotechnologies. Cambridge, MA: Blackwell.

Busch, L. W.B. Lacy, J. Burkhardt, D. Hemken, J. Moraga-Rojel, J. Souza Silva, and T. Koponen. 1995. Making Nature, Shaping Culture: Plant Biodiversity in Global Context. Lincoln: University of Nebraska Press.

Campbell, D. 1993. Ignorance kills co-ops. Michigan Milk Messenger. Oct/ Nov.: 26, 14.

Chesnick, D., and J. Staiert. 1992. Co-op involvement and opportunities in the seed industry. Agriculture Cooperative Service Research Report 105. Washington D.C.: U.S. Department of Agriculture.

Cheetham, P. S. J. 1993. The use of biotransformations for the production of flavors and fragrances. Trends in Biotechnology 11(11): 478-488.

Cleberg, H.D. 1995. The challenges and opportunities facing cooperatives. The Cooperator. Feb.: 6-7.

Coffey, J.D. 1993. Implications for farm supply cooperatives of the industrialization of agriculture. American Journal of Agricultural Economics 75(12): 1132-1136.

Collins, R.A. The conversion of cooperatives to publicly held corporations: A financial analysis of limited evidence. Western Journal of Agricultural Economics 16(2): 326-330.

Cropp, R., and G. Ingalsbe. 1989. Structure and scope of agricultural cooperatives. Cooperatives in Agriculture. In Cooperatives in Agriculture, D.W. Cobia (ed.). Englewood Cliffs, New Jersey: Prentice-Hall.

Gray, T.W. 1995. Rhetorical constructions and cooperative conversions: A comment. Paper presented at the annual meetings of the Rural Sociological Society, Washington, D.C., Aug. 17-20, 1995.

Hofstad, R. 1990. Tomorrow is today. American Cooperation 1990. Washington D.C.: National Council of Farmer Cooperatives, 9-12.

Hoiberg, E.O., and G. Bultena. 1990. Toward an assessment of the socioeconomic impacts of farmers' adoption of porcine growth hormone. In Agricultural Bioethics: Implications of Agricultural Biotechnology. Ames: Iowa State University Press, 1990.

Johnson, K.L. 1995. Contract cropping. Farm Industry News. Mid-February: 11.

Kibbe, T.F. 1995. The agricultural cooperative model and rural communities. Paper presented at the annual meetings of the Rural Sociological Society, Aug. 17-20, 1995.

Kidd, G., and J. Dvorak. 1995. A porter analysis of the corn-genetics industry. Bio/Technology 13(3): 211-212.

Kloppenburg, J.R., Jr. 1988. First the Seed: The Political Economy of Plant Biotechnology, 1492-2000. New York: Cambridge University Press.

Krimsky, S., and R. Wrubel. Forthcoming, 1996. Agricultural Biotechnology and the Environment: Science, Policy, and Social Issues. Urbana: University of Illinois Press.

Lacy, W.B., and L. Busch. 1988. Biotechnology: Consequences and strategies for cooperatives. In Biotechnology and Agricultural Cooperatives: Opportunities and Challenges, W. B. Lacy and L. Busch (eds). Lexington: Kentucky Agricultural Experiment Station.

Lacy, W.B., and L. Busch. 1989. Biotechnology: Challenge and opportunity for agricultural cooperatives. Policy Studies Journal 17: 203-214.

Law, B.A., and F. Mulholland. 1991. The influence of biotechnological developments on cheese manufacture. Biotechnology and Genetic Engineering Reviews 9: 369-409.

Lee, T.F. 1993. Gene future: The Promise and Perils of the New Biology. New York: Plenum Press.

Marion, B.W. 1986. The Organization and Performance of the U.S. Food System. Lexington, MA: Lexington Books.

NatureMark. 1995. NatureMark, background. Boise, ID: NatureMark. (Advertising Brochure).

Ratchford, C.B. 1990. Addendum to 'Biotechnology on the cutting edge.' American Cooperation 1990. Washington D.C.: National Council of Farmer Cooperatives, 34-35.

Reilly, J. 1991. An overview of the use of subsidiaries by agricultural cooperatives. Journal of Agricultural Taxation and Law 13(3): 197-235.

Schrader, L.F. 1989a. Equity capital and restructuring of cooperatives as investor-oriented firms. American Cooperation 1989. Washington D.C., National Council of Farmer Cooperatives, 41-53.

Schrader, L.F. 1989b. Future structure, problems, opportunities. In Cooperatives in Agriculture, D.W. Cobia (ed.). Englewood Cliffs, New Jersey, Prentice-Hall.

Schrader, L.F. 1990. Will success lead to a cooperative's demise? In American Cooperation 1990. Washington D.C.: National Council of Farmer Cooperatives, 13-19.

Seaman, W. 1993. Grain marketing and further processing: Making it work. In American Cooperation 1993. Washington D.C.: National Council of Farmer Cooperatives, 227-235.

Torgerson, R.E. 1993. Strategic alliances revisited: A cooperative perspective in American cooperation 1993. Washington D.C.: National Council of Farmer Cooperation, 151-59.

Torgerson, R.E. 1994. Co-op fever: Cooperative renaissance blooming on northern plains. Farmer Cooperatives 61(6): 12-14.

Urban, T.N. 1991. Agricultural industrialization: It's inevitable. Choices (Fourth Quarter): 4-6.

United States Department of Agriculture (USDA). 1986. Farmer Cooperative Statistics, 1986. Washington, D.C.: U.S. Department of Agriculture-Agricultural Cooperative Service, Service Report 19.

United States Department of Agriculture. 1992. Farmer cooperatives statistics, 1992. Washington D.C.: U.S. Department of Agriculture, Agricultural Cooperative Service, Service Report 39.

United States Department of Agriculture. 1993. Farmer cooperative statistics, 1993. Washington D.C.: U.S. Department of Agriculture-Agricultural Cooperative Service, Service Report 43.

Ward, C.E. 1993. How cooperatives meet the challenge of franchises. American Cooperation 1993. Washington D.C., National Council of Farmer Cooperation, 236-242.

Weaver, R.D., (ed.). 1993. U. S. Agricultural Research: Strategic Challenges and Options. Bethesda, MD: Agricultural Research Institute.

Young, B. 1991. The dairy industry: From yeomanry to the institutionalization of multilateral governance. In Governance of the American Economy, J. L. Campbell, J. R. Hollingsworth, and L. N. Lindberg (eds). Cambridge: Cambridge University Press.

6

Toward an Agriculture Information System To Maximize Value in Agricultural Data

Marie-Claude Fortin and Francis J. Pierce

There seems to be general agreement that the information age has reached the American farm. No consensus has emerged about what that statement means for agriculture as a whole or on the farm. What is clear is that the type and source of data and information generated and needed at the farm level are changing and the level of detail is becoming increasing complex. To some, this is not a positive development, for in their view, agriculture is already in information overload. The fact is, however, the rate of acquisition of information is rapidly increasing, currently doubling every year. Thus, the problem for the future is the increasing quantity of information since we lack the tools to manage it in ways that extract the desired value from a broad information base. Therefore, a pertinent topic in the discussion of the privatization of information and technology in agriculture is the development of an Agricultural Information System (AIS).

An excellent example of information-age technology currently on the farm is the application of site-specific or precision farming concepts. The goal here is to match inputs to localized conditions within a field. To achieve this, a farmer needs site-specific information about the conditions everywhere within a field and recommendations and application control technologies that optimize crop production for each condition. In this effort, farmers and their service providers are collecting data at levels of detail never before attempted, and when done correctly, at levels of quality never before achieved. For example, where farmers rarely have long-term yield records for a whole field, they now measure yield every second

ISBN 1-57444-104-3/98/$0.00/$.50

with yield monitors. For a 15-foot combine traveling at 3.5 mph, approximately 550 yield measurements are obtained per acre.

A current problem for farmers is their limited experience and expertise with such detailed data and its spatial format. Farmers will need to be able to collect, manage, and interpret data and information from disparate sources in order to extract value from it. In essence, the value of information-age technologies on the farm is found in the interpretation of data and information in its application to solving agronomic problems on the farm.

Precision farming is indicative of the information age for agriculture. There is increasingly a reversal of flow of data in agriculture, in which farmers who used to depend on public and private sources of information, are generating more pertinent and possibly better data on their farm than their traditional information providers. Farmers are also demanding more specific data and information to assist them in their farm management decisions, including digital and spatially referenced data rather than data and information in more traditional formats. In this regard, some are questioning the role of the public sector in providing information to farmers, particularly if their information is perceived to be inappropriate, in the wrong form, untimely, and/or unavailable. The fact that any deficiencies may be due to diminishing budgets and fewer personnel is not relevant to those who require the information. Little expertise has currently been developed to address farm-level data and information needs, and the public-private roles for delivery of this expertise to farmers is being worked out as technologies and usage evolve.

The farmer's goal is to turn their data into decisions for the future. The desire to make decisions from information and knowledge creates the need for an agriculture information system. An AIS will prove to be essential for extracting full value from information-age technologies on the farm. This paper presents a primer on the need for and essential components of an information system for agriculture.

OVERVIEW OF AN INFORMATION SYSTEM

Data and information are the foundation of all decisions. Data are transformed into information through organization and analysis, and ultimately into knowledge through the integration of multiple sources of information or through modeling (Figure 1). Knowledge and prediction influence decisions and their acquisition leads to empowerment. An information system is a tool for generating knowledge from digital data. In agriculture, great potential exists for full-capability information systems, i.e., information systems that combine efficient data man-

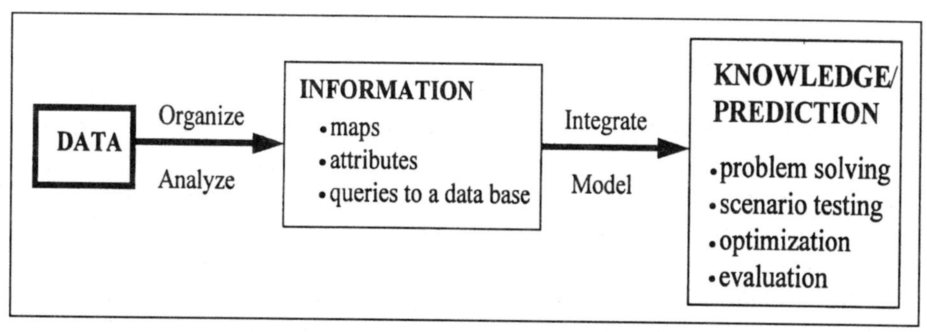

Figure 1. Information flow from the gathering of data to the acquisition of knowledge.

agement and which are flexible enough to integrate new data types or varied decision-making tools. Such information systems can be designed to manage data for a single farm or a whole country. More importantly, such information systems will allow exchange and transfer of data among users regardless of the size or scope of their operation.

The design of a computerized information system for agriculture can take many forms. Some systems focus heavily on the design of a database management system (DBMS). The main goal of a DBMS is to provide an environment that is both convenient and efficient to use in retrieving, updating, and storing information (Korth and Silberschatz 1991). Other systems concentrate more on the use of specific software tools or models (collectively termed decision support systems, DSSs) to solve problems. The data management component of a given DSS is usually specific to that application. A full-capability AIS combines a sophisticated DBMS, capable of data management for various applications with a variety of DSSs for problem solving. In addition, a full-capability AIS must be designed to incorporate advances in agricultural or software technology to avoid obsolescence. An AIS should have the capability to manage varying volume of data and assess data requirements for improvement of a desired end product. A full-capability AIS can provide decision-makers with solutions to data management, business, and scientific problems. While the design and implementation costs of a full-capability AIS are initially high compared to those systems designed to address specific problems, over time they become cost effective because full-capability AIS are generic systems that do not require redesign to resolve new problems. Market share in the information business will depend largely on the credibility, relevancy, and perceived value of an AIS.

Information systems are used successfully in business sectors, such as the banking, retailing, and airline industries. The agricultural sector will benefit from using similar information systems as information-age technologies increase in use and as information increasingly becomes a source of competitive advantage.

Agriculture utilizes a more diverse set of data than traditional businesses, including biophysical and land use data in addition to economics and social preferences data (Moon et al. 1995). A large part of the biophysical and land use data is spatially referenced and requires full integration with non-spatial or attribute data. In addition, sophisticated models are required for prediction of key processes such as crop growth, soil erosion, soil organic matter dynamics, and nutrient cycling. Therefore, the implementation of information system concepts must take in account the complexity of data and information in agriculture and their application to problem solving. In practice, a successful AIS can only be achieved by the collaboration of specialists from the domains of information systems, computer science, and agriculture.

There are several sound data management principles that form the basis for the design and development of an AIS. These principles need to be understood and applied rigorously to achieve a full-capability AIS.

- The first key to building a successful information system is the use of appropriate data management software.

The function of a Geographic Information Systems (GIS) is to facilitate the mapping and analysis of spatial data. Both spatial and attribute data must be managed within a GIS. Spatial data refer to features needed to generate maps and attribute data refer to characteristics, qualities or relationships of map features (Antenucci et al. 1991). In the application of a GIS, the first step is usually inventory followed by spatial analysis operations (Aronoff 1989). In this regard, the data management capabilities of GIS software are designed primarily to provide rapid retrieval and visual display of spatial data as maps and perform mathematical operations on a spatial data set. Consequently, data management capabilities of GIS software may not be efficient at data management and are frequently lacking in their ability to supply data to peripheral applications. Additionally, GIS packages do not effectively manage attribute data that are related to, but not part of, the spatial data base, for example, commodity prices. Thus, data management capabilities in GIS generally lack the full set of functionalities desired for the implementation of a full-capability AIS.

An AIS will require a database management software product in addition to a GIS or as a component of the GIS to manipulate the distribution, management, and storage of attribute data and to oversee the communications among all the components of the information system, including the GIS.

At present, a relational data base management system (RDBMS) constitutes the best choice for a data base management software. The database management system chosen to coordinate the various software tools of an agricultural business system must meet criteria of flexibility, performance, data integrity, and data security. Flexibility is key in an era of rapidly emerging technologies, exploding knowledge, and dynamic regulatory and organizational environments. Performance is key in an era of information-derived competitive advantage. Integrity/normalization is key in highly complex information environments, as AIS would surely be. Security is key as proprietary, commodified, and strategic aspects of information grow.

Flexibility and performance are usually found in top-of-the line software products. Corporations releasing these products also tend to create partnerships with leaders in other fields of the software industry (e.g., hardware vendors, application products, and more recently, GIS products). These partnerships usually result in compatibility of components or in the joint development of integrated system of components.

Two important issues in data management in an AIS are data integrity and data security (access to the data). Data integrity can be affected both in the quality of data collected and its management within the database itself. Poor quality data results from lack of quality control in sampling and analysis during data acquisition. For example, using sensors in precision farming, a farmer is able to collect data at high measurement frequencies. The quality of that data may be poor if the sensors are not properly installed, maintained, and calibrated. Given that farmers are now equipped with the technologies needed to evaluate products and services used on their farms, data quality issues emerge relative to liabilities associated with the perceived or real performance on the farm. Therefore, the data quality component of data integrity will be critical to both the farmer and agribusiness. Once entered, data integrity within the database is insured by using normalization rules within a relational database. Normalization, a technique well known by data management specialists, minimizes data redundancy. Data redundancy can cause integrity problems since update and delete transactions may not be consistently applied to all tables and all copies of the data. Normalization also helps identify missing links among tables or missing entities within a table (Oracle Corp. 1992).

Data security within an AIS is functionally attained through the application of strict database administration principles. In many systems, the data base administration function is assigned to one person who can, in turn, grant privileges for accessing certain parts of the data base or certain functions (view, enter, or update) to the other users based on their role within an organization. Data base administration establishes standards including definitions, quality, and timing. Proper database administration is essential in situations where multiple users will have access to a database.

Who has access and who controls farm-level data are emerging issues in the age of information. High measurement frequency data associated with precision farming can validate a farmer's production practices while revealing problems not previously identifiable. At the same time, if this level of detail is made available to others, it may reveal a farmer's production operation, placing them in jeopardy with those who would misuse it. In principle, farmers seem willing to share their data if its use by others provides value to their individual farming operation. Yet, farmers seem quite wary of those who plan to use their data in ways that may infringe upon their rights to privacy and self-determination. Data ownership and property rights are emerging issues of the information age that have no clear resolution, except to say that farmers own their data and retain rights to its power. Least clear in this issue is the ownership of data collected by others on an individual's property, such as a fertilizer dealer who provides a service to farmers or farmers who rent land and obtain data on that land in the process of managing it. However data ownership and use issues are resolved, data security is prerequisite if data ownership issues are to be properly accounted for within an AIS.

- The second key to a successful information system is the design and use of an efficient and flexible data model.

A data model is the formal definition of entities, their attributes and the relationships among entities in the database. It is the basis on which the various tables and the columns within each table of a database will be created. The data model will dictate in good part the performance of the information system. A data model provides the intelligence necessary to integrate different kinds of data and to coordinate the execution of various tools within the information system (Moon 1995). The successful data model for agriculture will allow for relatively easy integration of the database with varied tools, either commercially available software or in-house applications since changes to an information system are unavoidable due to modifications in agriculture production technology, in social and economic conditions, and in the constantly evolving needs of clients. The decision to use a new tool should not require a massive redesign of the AIS.

A data model must insure the scientific credibility of the data by providing a pedigree for each entry in the attribute or spatial database. Pedigrees can be implemented by requiring information about the entry (metadata) such as a date, a reference person or agency, and detailed methods of data collection. Validation checks can also be implemented by ensuring that specific types of data be within a certain range or else data is rejected.

For agricultural applications, a data model must also integrate spatial and non-spatial data. Spatial features such as lines and points or cells delineating fields or farms must be linked to their non-spatial attributes such as owner, soil type, crop type, soil depth etc... (Ulansky and Moon 1995). Linkage of GIS and other data base management software used to be executed by programmers. Some GIS software packages now offer direct links to specific DBMSs. The direct linkage of GIS to DBMS is an important consideration in selecting the component software technologies for an AIS.

- The third key to a successful information system is a friendly user interface.

A user interface should allow the end-user to utilize the powerful technologies of the information system to answer questions without requiring intimate knowledge of components, such as the DBMS or GIS (Moon et al. 1995). In practice, the user interface should allow for easy access to the functions of data entry, query, database improvement and to analytical procedures that execute the desired business functions (Moon et al. 1995).

AN EXAMPLE AIS

The status of agriculture and natural resources information and the needs of potential users of computerized systems have been recently addressed (Gardner 1993; Thompson and Weetman 1995; Lane 1995; Ross and Hannam 1995). Current systems consist mostly of decision-support technology designed for specific problems within specific regions. Few systems actually provide a full range of data management capabilities combined with decision-support technology. Fewer have actually been successfully implemented and tested in several different applications. Following is a description of an AIS which is unique in that it is generic (i.e., capable of integrating most decision-support systems software tools currently available with sophisticated data management). Its varied implementations since 1993 warrant a detailed description.

The Land Analysis and Decision Support System (LANDS) was developed from 1988 to 1993 using a multi-disciplinary team of nine person-years including agriculture specialists, information specialists, and programmers from the public and private sectors coordinated by the Research Branch of Agriculture and Agri-Food Canada. LANDS is used for agriculture and natural resources data management, applied problem solving and research. The original development of LANDS was funded by CIDA (Canadian International Development Agency) for the development of an information system for land use planning for Malaysia. Following the delivery of the system to the government of Malaysia in 1993, the LANDS original development team was replaced by a smaller maintenance team.

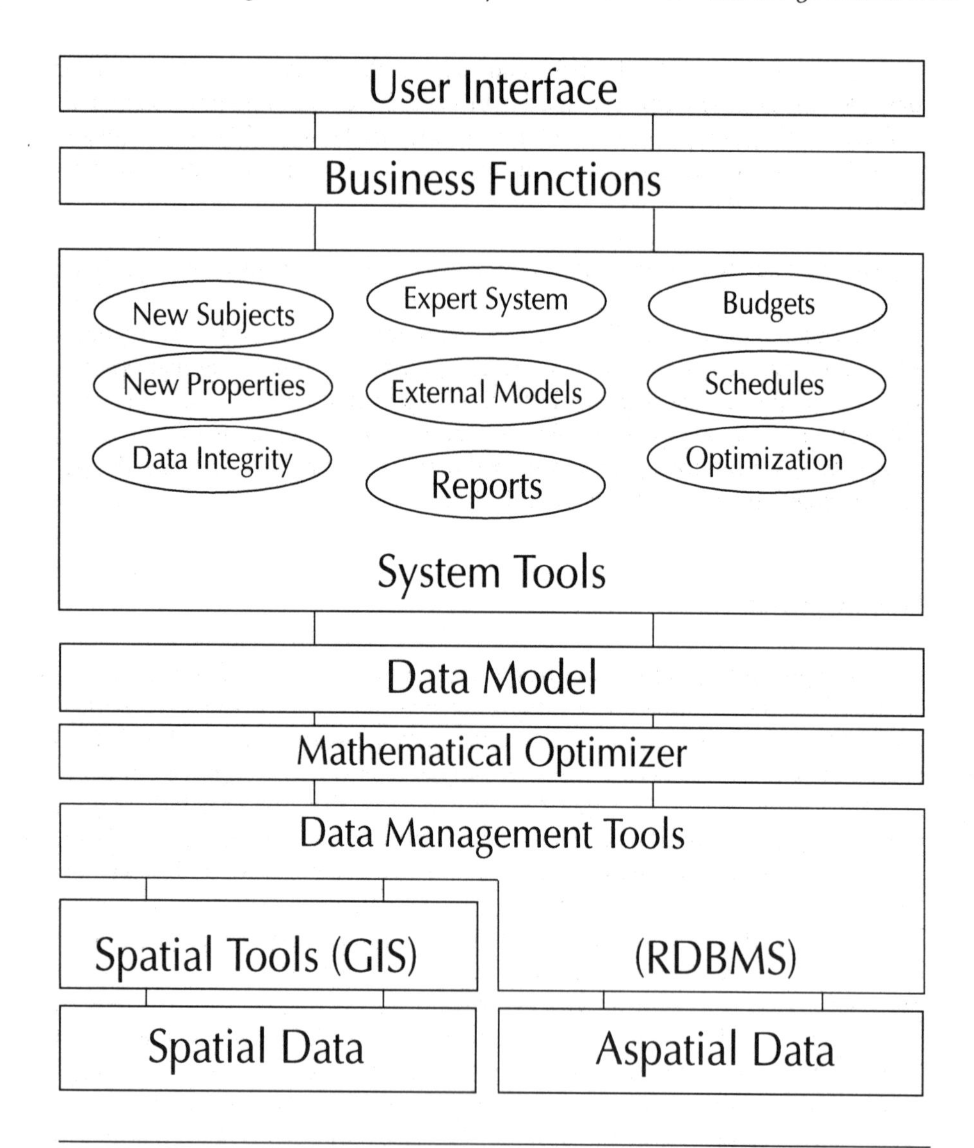

Figure 2. The structure of the LANDS agriculture information system is one of layered technologies. Spatial and non-spatial data constitute the basis of the information system. A geographic information system (GIS) is the organizer of spatial data and provides these data to the relational database management system (RDBMS) which allows integration of both types of data. In turn, the RDBMS provides data to another support technology, the linear optimizer. The data model defines how data is organized in tables and columns, and coordinates the execution of the various tools. Tools are for creating new entities and relationships, insuring integrity in the database, budgeting and scheduling, optimizing, reporting, using an expert system for matching entities with common elements, and for modeling. Business functions assemble the tools and the underlying technologies to answer specific questions. The user-interface provides access to this set of technologies.

In North America, the LANDS information system has been adopted by two Canadian federal offices responsible for natural resources monitoring and by a multi-disciplinary university research team for impact assessment of agriculture on health. The applicability of the information system to forest management is being reviewed in connection with a joint project involving the private sector and Agriculture and Agri-Food Canada. Since 1993, LANDS has also been used to solve specific problems for various clients including evaluation of the economic impact of conservation tillage in the northern Canadian Prairies, establishment of a GIS-based information system for irrigation monitoring and water use prediction, and for evaluation of nitrate pollution risks associated with various animal and crop production systems over a British Columbia-Washington state aquifer. For the latter project, a series of optimized solutions to reduce nitrate levels in the aquifer to acceptable levels were presented to decision-makers. These solutions were calculated using the results of modeling various waste management options.

The structure of the LANDS information system is one of layered technologies (Figure 2) (Moon et al. 1995). At the base of the agriculture business system lies non-spatial or attribute data and spatial data. As suggested earlier, the two essential technologies managing and organizing data and providing necessary inputs to the tools listed above consist of a relational database management system (RDBMS) for attribute data and a GIS for spatial data. The GIS plays an essential but minor role as a spatial data organizer and provider to the database management system (Moon 1995). Conversely, the RDBMS allows for the integration of spatial and attribute data, and must provide data to other support technologies which include a linear optimizer. The linear optimizer is a mathematical procedure for determining the optimal allocation of limited resources. The data model coordinates the execution of the various system tools. The system includes tools for creating new entities and relationships and insuring data integrity, a budgeting and scheduling application, an optimization tool, a report tool, and an expert system or rule-based logic for matching subjects with common elements (e.g., to assess suitability of land for a specific purpose such as waste disposal). In addition, various models have been added to the tools to satisfy the needs of specific clients. Business management functions assemble the tools and the enabling technologies into a system that answers the specific questions of the clients (Moon et al. 1995). Finally, the LANDS user-interface allows a user without specialized skills to access the powerful technologies integrated within the information system. The layered technology approach has succeeded in creating a versatile and time-tested information system.

SUMMARY

The value of information is found in its use. Our contention is that an information system is necessary to effectively extract the desired value of data and information in production agriculture. The need for an AIS is becoming increasingly urgent given the proliferation of information-age technologies on the farm. For farmers, the value of information-age technologies is in solving problems and validating their management practices. The discussion of an AIS is very relevant to the overall discussion of the privatization of information and technology in agriculture. Who has access and who controls farm-level data may have important ramifications for the structure of production agriculture in the 21st century, particularly in farm size, the need for outside services on the farm, and the flow of goods and services to and from the farm. If properly constructed, an AIS should link the public and private sectors in their efforts to serve the needs of the farmer while at the same time provide benefits for both sectors. An AIS provides controlled and efficient access to information. As described in this paper, such a system will facilitate problem solving and innovation while protecting data integrity and security for those who own the data. The design and development of an AIS should be a priority for agriculture. Fortunately, information systems are not new, and their proven utility in other business sectors should provide a model for their development in agriculture.

REFERENCES

Antenucci, J.C., K. Brown, P.L. Croswell, M.J. Kevany, and H. Archer. 1991. Geographic Information Systems: A Guide to the Technology. New York, NY: Van Nostrand Reinhold.

Aronoff, S. 1989. Geographic Information Systems. A Management Perspective. Ottawa, ON, Canada: WDL Publications.

Gardner, W.R. 1993. A call to action. Soil Science Society of America Journal, 57:1403-1405.

Korth, H.F., and A. Silberschatz. 1991. Database System Concepts. New York, NY: McGraw-Hill.

Lane, L.J. 1995. Natural resources management. The emergence of multi-objective decision support systems. Preconference Proceedings, First International Conference on multiple objective decision support systems for land, water and environmental management: concepts, approaches and applications. Univ. Hawaii, Honolulu, HI. July 23-28.

Moon, D.E. 1995. GIS in Agriculture. 50th Annual Regional Conference, Pacific Northwest Section of the ASAE, Oct. 1-3, Harrison Hot Springs, BC, Canada.

Moon, D.E., S.C. Jeck, and C.J. Selby. 1995. Elements of a multiple goal DSS: Information, model and user management. First International Conference on Multiple Objective Decision Support Systems, July 23-28, Honolulu, HI.

Oracle Corp., 1992. Data modelling and database design. Vol.1. Participant Guide. Oracle Corporation, Redwood Shores, CA.

Ross, J., and I. Hannam. 1995. Multi-objective decision support systems used in the management of forest ecosystems in south east Australia. Preconference Proceedings, First International Conference on multiple objective decision support systems for land, water and environmental management: concepts, approaches and applications. Univ. Hawaii, Honolulu, HI. July 23-28.

Thompson, W.A., and G.F. Weetman. 1995. Decision support systems for silviculture planning in Canada. The Forestry Chronicle 71(3):291-298.

Ulansky, S.M., and D.E. Moon. 1994. Compilation of disparate digital spatial data. Agronomy Abstracts, American Society of Agronomy, 86th Annual Meetings, Seattle, WA.

7

The Political Economy of Precision Farming

Steven A. Wolf and Frederick H. Buttel

Precision farming (also known as variable-rate, prescription, site-specific, or soil-specific farming) is rapidly becoming the most touted of emerging technologies in American agriculture. Scarcely an issue of the mainstream agricultural print media is now published without at least some mention of the current applications of and new developments in precision farming. Precision farming—application of georeferencing technology (i.e., global positioning satellite system [GPS]) for purposes of data recording and navigational control for field operations—is increasingly hailed as a means of increasing production efficiency while simultaneously mitigating agricultural pollution. The claims surrounding precision farming—that capitalizing on extant agroecological heterogeneity, through a just-enough and just-in-time approach, will produce economic and environmental benefits simultaneously—are clearly very appealing. Further, we think it is very likely that precision farming and its variants will play a growing role in the restructuring of agricultural practices over the next several decades.

Unlike other emerging agricultural technologies, particularly biotechnologies, precision farming has received relatively little attention from the social sciences. To the degree that there has been attention to precision farming technology from a social science point of view, it has focused mainly on the credibility of the profitability and environmental claims made for the technology. Interestingly,

This paper was originally published in the American Journal of Agricultural Economics Vol. 78. (December 1996) and is reprinted with permission of the American Agricultural Economics Association.

ISBN 1-57444-104-3/98/$0.00/$.50

much of this scant social science and related literature has reported that there is no clear evidence for these economic and environmental claims. This is not to say that there will not be economic and environmental benefits to the technology when it is further developed and refined. Nonetheless, the lack of evidence for some of the most critical claims for precision farming technology suggests that we should look beyond the private profitability and environmental rationales in order to understand the significance of precision farming.

THE SOCIOECONOMIC DISTINCTIVENESS OF PRECISION FARMING TECHNOLOGY

Precision farming is not only an important new technology in its own right, but its development and commercialization are compelling and revealing in many respects. First, the vague alignment of the claims made for and the existing evidence on precision farming suggest the need to eschew a straightforward induced-innovation (or "demand" for science) understanding of its R&D, and thus to give attention to the scientific/agribusiness promotional ("supply" of science) perspective as well.[1] Indeed, the commercialization of these technologies thus has had as much to do with promotion and the search for new uses of existing non-agribusiness manufacturing and/or military technology as it has had to do with the new technology being induced by economic or social signals. This pattern of technology transfer is consistent with the U.S. Environmental Protection Agency's Environmental Technology Initiative, a competitive grant program that explicitly seeks to convert defense technologies to environmental applications.

Second, precision farming is being developed and deployed in an era of decline of public sector agricultural research, extension, and information transfer across the globe (von Braun et al. 1993; Eicher 1994). This decline in public institutional authority is occurring at a time when information is being explicitly identified as a source of competitive advantage and is therefore increasingly internalized by firms. As precision farming is designed to support continuous production monitoring and thus improvement over time, it both reflects and propels the resulting trends of privatization of agricultural technology and information transfer.

Third, precision farming is largely a mechanization technology. Specifically, precision farming combines mechanization as commonly understood (substitution of inanimate energy for human labor) with new forms of mechanization (information generation and analysis, through computerized GPS, GIS, sensors, and on-the-go yield monitors (Cochrane, 1979)). Much like previous mechanization technologies, one can expect that precision farming will be characterized by capital intensity and significant scale economies, and in the long term will contribute to the concentration of agricultural production in fewer hands. In this sense precision farming is an industrializing technology as it is consistent with a pattern of "appropriation and substitution" (Goodman et al. 1987) through which the significance of atomistic farmers' contribution to the production process has been marginalized relative to that of centralized agribusiness. According to Krimsky and Wrubel (1996:19), appropriation and substitution appear to represent "an important selection pressure for agricultural innovation before questions of efficacy, economic efficiency, and markets are considered."

Finally, precision farming is essentially premised on a continuing trend toward relatively inexpensive chemical inputs. The effectiveness of precision farming is based as much on applying greater amounts of chemical inputs to the soils that exhibit profitable responses to them as it is on reducing chemical inputs in less productive or environmentally sensitive areas. Insofar as the technology will maintain, if not increase, the use of chemicals, its utility will be shaped by the relative costs of these inputs and the degree to which farming systems dependent on these inputs remain competitive.

AGRICULTURAL TECHNOLOGY AND THE MANAGEMENT OF AGROECOLOGICAL DIVERSITY

Unlike the industrialist, who formulates management strategy with a presumption that essentially the same type of technology will be employed regardless of the location of production, for the agriculturist many of his or her most critical management decisions will be about how to select among technologies in order to deal with the biophysical diversity that exists across fields and landscapes. Indeed, the biological nature of agriculture and its agroecological diversity make homogenization of production practices highly problematic, if not impossible (Goodman et al. 1987). All farming systems are by necessity based on particular strategies for management of agroecological diversity.

Traditional agricultural production systems, typified by many (but not all) peasant agricultures, have typically involved the employment of agroecological, biological, and genetic diversity as management tools to accomplish goals such as food sufficiency and minimization of risks associated with variables such as pest outbreaks and weather. Selection among species and cultivars was employed to enhance performance, though within an overall framework of maintenance of species and cultivar diversity and agroecosystem resilience. Such traditional peasant production systems, which largely dominated worldwide until the late nineteenth century, have largely disappeared a century later.

Driven by the availability of a small cluster of complementary generic technologies (chiefly chemical fertilizers, synthetic pesticides, large scale tractor-based mechanization, and fertilizer-responsive, higher-yielding, genetically uniform crop varieties), world agricultures have been dramatically transformed. The essential feature of this world-agricultural transformation—which we refer to as the "industrial intensification of agriculture" (IIA)—has been that IIA technologies have enabled researchers and agriculturalists to override or suppress substantially agroecological diversity and thereby routinize production through monocultural production practices and enterprise specialization. The world-scale employment of the four generic IIA agricultural production technologies has historically had a number of positive outcomes. Total factor productivity and output have increased, and relative food prices have decreased (Lipton with Longhurst 1989). Employment of these four generic technologies was also an understandable strategy given the long history of underinvestment in public agricultural research, since limited public research funds could be allocated most effectively through a focus on fine-tuning a restricted repertoire of technologies.

But despite the successes registered by world agricultures after World War II, the increasingly intensive use of the four generic technologies has directly or indirectly led to several significant problems. These have included environmental problems (chiefly pollution due to surface and subsurface run-off from fields and feedlots, land degradation, and disruption of ecological structures and functions), the rising incidence of pest resistance to biocides, certain inefficiencies (e.g., prophylactic use of biocides, waste of animal manures), market barriers reflected in saturation of demand for basic food commodities, and the threat that the generic twentieth-century agricultural technology package would begin to reach diminishing returns to inputs. These and related concerns have spurred the attention of many scientists and policymakers over the past 25 or so years, and, more recently, the attention of agribusinesses. In many instances, these responses were catalyzed by the efforts of social movements across the globe that aim to enhance agriculture's environmental performance and alter the trajectories of agricultural development unleashed since World War II.

These movements have sought to block further industrialization of agriculture, develop alternative-sustainable agricultural systems and, in some countries and regions, preserve traditional peasant agricultures. Among these movements, definitions of sustainability have been quite diverse. Some quarters of the movement have aimed at radical changes in institutions, while others have stressed the need for radical shifts in the nature of technologies (e.g., Wes Jackson's advocacy of permaculture). But, in the main, the alternative systems that have been most successful in gaining acceptance and credibility are those that are relatively compatible with the structural changes in world agricultures that have been induced by IIA-namely, predominance of large, highly capitalized family-proprietor farms, specialization of farms and regions, and increased contractual integration with non-farm agribusinesses. Thus, incremental efforts to enhance the environmental performance of "mainstream" agriculture, while retaining its output and productivity advantages, have tended to predominate over more radical visions of sustainable agriculture.

PRECISION FARMING WITHIN AN INDUSTRIALIZING AGRICULTURE

To this point we have expressed precision farming, in part, as a defensive or conservative reaction of an agricultural system struggling to maintain legitimacy in the face of mounting social, economic and ecological challenges. We now expand the scope of our analysis, and return to our initial premise that a "science-push perspective" is instructive as applied to precision farming, by considering how precision farming advances the interests of non-farm agribusinesses within a rapidly restructuring agriculture.

Coordination of production

At the field- and farm-level precision farming is a production and planning tool supporting more refined application of inputs and better investment decisions. At the level of the larger production subsector, precision farming can be considered a coordinating technology providing a digital interface between the farm field and other stages of production. Precision farming is emerging at a time when coordination within and between firms has assumed a new level of importance (Streeter et al. 1991). Factors contributing to this focus include structural change in farm and agribusiness sectors and availability of new computer-based technologies. Additional considerations include chronically depressed commodity markets, vertical coordination of commodity subsectors, differentiation of consumer products, and heightened concern applied to product liability. These last four interrelated factors are discussed briefly below.

Surplus production capacity, global sourcing of increasingly interchangeable raw materials, and development of industrially manufactured substitutes for products previously produced through biological processes serve to maintain pressure on individual firms and commodity subsectors to enhance productivity. The search for precision is a search for competitive advantage in production efficiency. Productivity gains are sought in field operations, for example fertilize on an as-needed basis, but also through tighter choreography of activity within multi-institutional, coordinated systems of manufacture. Detailed and geographically and temporally specific data are a powerful means to analyze and schedule logistical elements of complex, capital intense production systems. Because precision farming data are digital, they can be communicated up and down the commodity chain in real-time and can be easily translated into a format compatible with the operating systems of multiple users as required in coordinated systems of manufacture.

In order to react to rapidly shifting consumer preferences and take advantage of improvements in materials handling and processing facilities design, processors are now seeking to expand control over the character of their feedstocks. Reportedly, commodity markets for raw, fungible goods are being replaced by markets for specialized, identity-preserved products where attributes such as size, color, moisture, starch, and protein content of goods are specified. Developing informational attributes allows firms to disaggregate their goods from those of competitors and attract premium prices. The ability to more closely regulate crop inputs (e.g., seed, water, fertilizer, pesticides, growth regulators) and maintain detailed, readily accessible production records through precision technologies supports interest of many parties in production systems who benefit from developing, maintaining, and augmenting product attributes desired in the marketplace.

Coordinated systems of manufacture and product differentiation demand accountability. As firms seek to preserve crop attributes, "brand" their products, or engage in other value-added strategies, concern over product and production liability increases. Further, the implications of loss of consumer confidence motivates firms to seek out suppliers and purchasers who are capable of comprehensively documenting their activities. As historic examples (e.g., alar) and contemporary concern over contamination of berries in U.S. markets illustrate, food safety concerns are a powerful reason for firms to develop quality controls, monitor themselves, and have ready access to detailed information about the origin and destination of goods with which they have contact. Parallel to concerns over food safety, the ability to track pesticides and fertilizers from the point of manufacture to the farm field—life cycle accounting—has taken on new significance. Analysis of claims applied to worker protection, environmental contamination, pesticide drift, and product performance guarantees will be aided by geographically and temporally specific data. Precision farming is compatible with and is an extension of highly capitalized firms' growing efforts to control their operating environments in order to manage risks.

Product differentiation, just-in-time production, vertical coordination, production monitoring, and precision farming represent a set of convergent institutional and technological innovations. Precision farming potentially will contribute to maintenance of economically competitive production systems in globalizing markets, support preservation and enhancement of desired crop attributes, facilitate communication within increasingly coordinated production systems, and promote accountability as applied to agrichemical use. In an era in which farms are increasingly part of the larger agricultural production, processing, and marketing process, precision farming mechanizes on-farm information generation and analysis consistent with information management developments in off-farm segments of commodity chains. From the perspective of the overall production process, precision farming avoids a discontinuity and potential bottleneck in the flow of information up and down commodity chains.

Maturation of crop input markets

Competition within a mature agrichemical industry is fierce in many regions and restructuring and consolidation are proceeding rapidly. Within this context, precision farming is simultaneously a means for local agribusinesses—fertilizer and pesticide retailers—to differentiate themselves from local competitors, add value to their products and diversify their businesses through fee-based service offerings (Wolf and Nowak 1995). Provision of service support will also serve to solidify relationships between dealers and farmers. Due to the scale, technical complexity, and infrastructural requirements of precision farming, the vast majority of farmers applying the technology will rely on off-farm service support. Soil testing, fertility mapping, custom application, component calibration, data analysis, data archieving, and report preparation are among the many spin-off services that will serve to integrate dealers into field- and farm-level activity.

At a time when "middlemen" in many industries are experiencing marginalization and potential elimination, precision farming can be seen as an opportunity for local retail dealers to secure a role in rapidly restructuring agricultural production systems. The expanding presence of regional agrichemical wholesalers and multinational manufacturers in community-based retail markets is making it difficult for some retail dealers to compete on the basis of price. Additionally, in some locations the growing presence of independent crop consultants is eroding retailers' status as a primary source of input management recommendations, further weakening their position within farming systems.[2] Local dealers of seed, fertilizer, and pesticides are clearly served by a technology which demands farmers engage off-farm service firms, especially a technology built around these very same inputs.

Beyond raising the significance of dealers' contribution within farming systems, precision farming confers opportunities upon local agribusiness as they obtain a degree of control over subscribing farmers' input management and resource allocation. Exercising a leadership role in customers' operations makes it possible to practice proactive management rather than being forced to react to spatially and temporally uncoordinated demand for pest scouting, soil testing, product delivery, and custom application services. By participating in the selection and scheduling of field operations, precision farming will allow dealers to manage more precisely their chemical inventory, physical plant, machinery, and labor resources.

The previous set of remarks address the interests of local agribusinesses, firms important to the marketing and on-the-ground implementation of precision farming. At the level of industrial agribusiness (e.g., input and machinery manufacturers), precision farming is attractive in that it provides access to farm- and field-level production activities. This development has significance in that there is currently a growing struggle amongst input manufacturers as to which inputs, in what proportions, will capture farmers' consumer purchases. At this time there is growth in the extent to which component production factors—e.g., fertilizer, pesticides, seeds, machinery, computers, and data—can be substituted for one another. For example, purchase of more expensive, herbicide tolerant seeds may lead to fewer or cheaper herbicide applications. Or, in the case of precision farming, investment in management information systems may reduce agrichemical expenditures. As the number of potential factor combinations rises and input markets continue to consolidate, it is increasingly attractive for manufacturers to be positioned to exercise leadership inside the farm gate. The development, packaging and sale of information and information technology by agribusiness and the reciprocal consumption by farmers represents a pattern of new and strengthened linkages between farmers and off-farm firms. This pattern of linkages indicates an accelerated penetration of field- and farm-level management by industrial capital and a transformation of the social relations of agricultural production (Wolf and Wood forthcoming).

Ongoing changes indicate that the interests of local suppliers and multinational manufacturers of agrichemicals and other inputs are increasingly consistent. An important feature of this coincidence of interests is that of gaining access to on-farm management processes. The ongoing blurring of traditional levels in pesticide and fertilizer marketing channels—manufacturers, wholesalers, distributors, and retailers—is illustrated by tighter relations between farmers and input manufacturers (e.g., direct marketing to farmers and increasingly sophisticated consulting services provided by manufacturers' representatives) and growing linkages between manufacturers and farm suppliers (e.g., strategic alliances, vertical integration). In an era of structural and technological change and potential volatility in the relative cost of factors of production, cementing relationships with end users of IIA technologies is consistent with the interests of both local and industrial agribusiness.

CONCLUSION

It is within the context of structural transformation and political-environmental scrutiny of modern agriculture that precision farming has leapt onto the scene. The technology reflects, and its future fine-tuning and applications must overcome, a crucial contradiction: On one hand, precision farming is a technical form that involves acknowledgment of the fundamental significance of biophysical heterogeneity. On the other hand, through incorporation of heterogeneity, precision farming essentially preserves the role of core IIA inputs—chemical fertilizers, synthetic pesticides, mechanization, and fertilizer-responsive crop varieties—and many of the practices, such as large-scale monoculture, that have been closely associated with their use. Precision farming is thus, somewhat ironically, a high-technology means for rationalizing the use of old—if not obsolete—farming inputs. It can be seen as part of a process of restructuring of agricultural practices so as to redress a set of problems caused by conventional IIA practices, while simultaneously protecting and advancing the industrial structures, investments, and institutional arrangements premised on these practices.

NOTES

[1] Here we follow Busch et al. (1991), who have argued that the conventional view of science and technology as demand-driven (R&D as a response to social needs and relative factor prices) needs to be balanced with a supply-driven perspective on science and technology. A supply-driven perspective is one that emphasizes that scientists, scientific institutions, science policy-makers, and technology developers have an interest in defining social problems as being scientific and technical ones, and in promoting science in general, and often particular technologies, as solutions to these problems.

[2] The advent and explosive popularity among agrichemical dealers of the American Society of Agronomy administered Certified Crop Advisory (CCA) program is an additional indication of these local firms' interest in increasing their presence inside the farm gate. Also, dealers increasingly providing no-interest operating loans to farmers indicates the competitiveness in the industry and their need to "lock in" their customers.

REFERENCES

Busch, L., W. Lacy, J. Burkhardt, and L. Lacy. 1991. Plants, Power and Profit. Oxford: Basil Blackwell.

Cochrane, W. 1979. The Development of American Agriculture. Minneapolis: University of Minnesota Press.

Eicher, C. 1994. Building productive national and international agriculture research systems. In Agriculture, Environment and Health, V. Ruttan (ed.). Minneapolis: University of Minnesota Press.

Goodman, D., B. Sorj, and J. Wilkinson. 1987. From Farming to Biotechnology. Oxford: Basil Blackwell.

Gorman, P. 1995. The paper presented swords into plowshares. Paper presented at the Power and Politics of Information. Center for Agriculture and Rural Development National Forum for Agriculture. Des Moines.

Krimskey , S., and R. Wrubel. 1996. Agricultural Biotechnology and the Environment. Urbana: University of Illinois Press.

Lipton, M., with R. Longhurst. 1989. New Seeds and Poor People. Baltimore: Johns Hopkins Press.

Streeter, D., S. Sonka, M. Hudson. 1991. Information technology, coordination and competitiveness in the food and agribusiness sector. American Journal of Agricultural Economics, 73 (December 1991)1465-1471.

von Braun, J., R. Hopkins, D. Puete, and R. Pandye-Lorch. 1993. Aid to Agriculture: Reversing the Decline. Washington D.C.: International Food Policy Research Institute.

Wolf, S., and P. Nowak. 1995. Development of information intensive agrichemical management services in Wisconsin. Environmental Management. 19(3):371-382.

Wolf, S., and S. Wood. Forthcoming. Precision farming: Environmental legitimation, commodification of information, and industrial coordination. Rural Sociology.

8

An Industrializing Animal Agriculture: Challenges and Opportunities Associated with Clustering

Amy Purvis

In the U.S. livestock and poultry sectors, bigger and fewer farms are producing increasing quantities of meat, milk, and eggs. Animal agriculture is be coming industrialized, as defined by Rhodes (1995): production occurs in specialized facilities tended by specialized labor using routine methods. Increasingly, producers tend to focus on animal husbandry, purchasing feedstuffs rather than integrating crop and livestock production activities at the same site. As industrialization trends accelerate, crop and livestock production are often spatially separate enterprises, thus surplus nutrients from manure sometimes accumulate resulting in potential problems with odor and water pollution. These environmental spillovers become an issue, both for producers and their neighbors, especially as animal agriculture facilities get larger and locate in clusters near processing facilities and specialized infrastructure amenities.

Poultry, cattle-feeding, swine, dairy, and cattle-ranching are at different stages along the industrialization continuum. To coordinate production and marketing functions, each animal agriculture sector has developed an evolving web of vertical and horizontal relationships (within and between firms).[1]

Helpful conversations and comments from Charlie Abdalla, Richard Conner, John Holt, Les Lanyon, Larry Libby, Peter Kuch, Jim McGrann, Fran Pierce, Kitty Smith, and Steve Wolf are gratefully acknowledged. Nicole Elmer provided valuable research assistance.

This research was performed in conjunction with a collaborative research project between Penn State University and Texas A&M University sponsored by the USDA/CSRC National Research Initiative (#9404099).

ISBN 1-57444-104-3/98/$0.00/$.50

Economies of size in production and processing technologies—as well as lower transaction costs from coordinating feedstuff-procurement, meat-packing and production activities—are the major economic forces driving industrialization in animal agriculture. Thus the trend favoring larger and fewer firms is likely to continue. As animal agriculture sectors become more concentrated, firms are increasingly likely to hoard rather than share strategic information. The fundamental premise of this chapter is that in the framing of a research agenda about privatization trends and industrializing animal agriculture, the public-private balance in decision-making about environmental management is a central consideration.

The relationship between industrialization and the public-private information balance is complex. Expansion and consolidation in animal agriculture have environmental side-effects. Environmental compliance and appeasing neighbors are crucial considerations in long-run decisions about location of facilities and the configuration of production systems, as well as in short-run production management decisions. Increasingly, both manure management experts and law/policy experts (Van Horn 1995; Merrill 1995a) predict that the economic viability or obsolescence of a livestock or poultry production facility are likely to hinge on the satisfactory performance of water protection technologies and success in controlling nuisance odor. Where animal agriculture is expanding, communities welcome the economic growth but with a caveat: environmental protection cannot be ignored. Neighbors in close proximity to expanding facilities voice the loudest concerns. As size and concentration of facilities increase, protests become more shrill.

The central thesis of this chapter is two-fold. Scenarios can be constructed where it would be mutually advantageous for animal agriculture and environmental regulators to work together in crafting environmental performance standards and policy enforcement mechanisms to promote responsible stewardship of water and air resources, as well as in coordinating the technological innovation and information dissemination necessary to achieve environmental protection most effectively. On the other hand, without purposive orchestration and adaptive management (Lee 1993), the current incentives work against producers sharing their cutting-edge environmental management data and technologies. These issues are inextricably linked with overarching policy issues concerning privatization of technology and information: while the environmental side effects of animal agriculture are viewed in localized areas as severe and getting worse, the current mix of policies and institutions do not support data-sharing and collaboration among producers and regulators.

In order to understand the obstacles and opportunities for improved environmental stewardship as a by-product of industrialization of animal agriculture, better information and data is needed concerning firm-level decision making and the dynamically-changing industrial organization of animal agriculture. To lay the groundwork for a policy-oriented empirical research agenda in this arena, the main task of this chapter is a description of the current industrial organization of the cattle-feeding, poultry, swine, and dairy industries. This historical and descriptive profile emphasizes the importance of regional shifts over the past 40 years in the consolidation and concentration of animal agriculture sectors. An understanding of how animal agriculture operates today and how it arrived at its current structure and character is offered as a necessary foundation for further inquiry and innovation regarding how to improve information flows and institutional arrangements in order to make environmental compliance policies perform better. The concluding section of this chapter proposes three themes as the starting place for a research agenda aimed at informing the design and implementation of workable and sustainable environmental policies for an industrializing animal agriculture.

ENVIRONMENTAL ISSUES FACING AN INDUSTRIALIZING ANIMAL AGRICULTURE

Both producers and rural residents have an economic stake in deciding what constitutes environmental compliance and how much is enough. Rural residents perceive that proximity to large-scale animal production will exert downward pressure on the value of their land and homes, perceptions reinforced by preliminary empirical evidence (Palmquist et al. 1995).[2] From producers' perspective, livestock and poultry facilities involve large capital expenditures with significant sunk costs (asset fixity). For them, ongoing lawsuits and public hearings—resulting in environmental compliance requirements remaining a moving target over months and years—are more than an expense, more than transaction costs (Stalcup 1993; Merrill 1995b). Producers face persistent worries about whether they have a future in a given location. In the short run, therefore, they are reluctant to increase their fixed investment which, accordingly, dampens innovation (Pagano 1993). Instead, they are likely to continue operating with an existing site and technology until they can determine whether they have a future—and at what cost—or whether they might better change locations.

It is an oversimplification to assert that livestock and poultry producers change locations in order to escape stringent environmental regulations.[3] Factors motivating recent regional shifts in animal agriculture have been motivated by a complex mosaic of marketing, production, and environmental policy factors (Abdalla et al. 1995). Producers and consumers negotiating mutually-acceptable environmental protection standards and their enforcement face an ad hoc maze involving a patchwork of federal, state, and local laws and procedures (Smith and Kuch 1995). Generally speaking, both communities and livestock producers have been left disappointed because their pacts about environmental compliance have been neither lasting nor satisfactory. Few if any communities with recently established clusters of animal agriculture production facilities can boast of a stable equilibrium balancing sustainable environmental and economic development goals. Because of differences in soils, climates, and communities' preferences, it makes sense that states have been delegated authority to enforce federal standards for confined animal feeding operations (CAFOs), as is the case with the 1972 Clean Water Act provisions[4] and the recently re-authorized Coastal Zone Management Act guidelines. These policy trends, however, promise little improvement for reducing the current environmental policy fragmentation, or improving the prospects and capacity for communities and industrializing producers striving to coexist in harmony.

It is the point of departure for this chapter that an uncoordinated and evolving set of environmental compliance policies—increasingly implemented at the local level on a case-by-case basis—serves neither the interests of an industrializing animal agriculture nor of the communities where facilities are established or relocating. Perpetual relocation is an expensive process which leaves in its wake hard feelings on both sides and, potentially, environmental and ecological disturbances due to mismanagement of nutrients polluting groundwater or surface water. Producers and communities both stand to gain from an emergent and stable consensus on what constitutes environmental compliance in animal agriculture and how to achieve it. Currently, a major obstacle to consensus is an inadequate scientific understanding of the cause-and-effect relationships between technology, management, and the resulting quality of air and water. Improved understanding of causal linkages and associated technological innovation is most likely to flourish in a situation where producers are motivated to openly exchange environmental monitoring data and information about what works best to minimize environmental effects.

Animal production facilities clustering in close proximity and sectors becoming consolidated and vertically integrated are at once an opportunity and a constraint to an open and iterative process for crafting fair and effective environmental policies based on multilateral communication between producers and regulators and affected communities. On the one hand, clustering would seem to afford opportunities both for monitoring water and air quality cost-effectively and for establishing improved information-exchange networks. Industrializing animal agriculture has a stake in more open, more predictable, and less capricious environmental regulations, which might be facilitated by collaborative learning between producers and regulators through expanded environmental monitoring and information networking. On the other hand, the trend toward increased concentration and specialization in production and processing is likely to motivate greater competition, thus more strategic and internalized information will be held closely by individual producers and firms. Further, in tandem with specialization, an expanding market for the privatization and internalization of environmental management functions is likely to develop, whereby cutting-edge solutions are increasingly viewed as sources of comparative advantage rather than as shared resources.

The future shape of animal agriculture will be conditioned by an evolving policy context for communities' and firms' decision-making, as well as by its past. In the next section, descriptions are presented of historical trends and environmental issues facing the beef cattle sector, the poultry sector, the swine sector, and the dairy sector. Poultry and cattle-feeding are the two most concentrated sectors in animal agriculture but they are organized quite differently. Major transformations in poultry and cattle-feeding have taken place over the past 40 to 50 years but current configurations of these sectors have been relatively stable in the 1980s and 1990s. The swine industry is undergoing rapid change in the 1990s. Astute observers ask whether the hog production and processing industries are more likely to follow the organizational model of the cattle-feeding sector or the poultry sector, or (more likely) to develop some organizational hybrid all its own (Rhodes 1995). The dairy industry has experienced significant regional shifts over the past decade largely driven by technological innovations. Attempts to understand rapid changes in vertical and horizontal relationships in animal agriculture have prompted the observation that the industry's structure "is evolving faster in practice than the theories we have to explain them" (Westgren 1994).

CATTLE FEEDING

Historical trends

A cluster of feedlots was established in the High Plains (the Panhandle region of Texas, Oklahoma, and New Mexico) and grew exponentially in the 1960s and 1970s (Dietrich et al. 1985). Brown (1993) offered a profile of the cattle-feeding sector and its beginnings: in 1960, the Corn Belt and the Central Plains produced 56 percent of fed cattle and the High Plains produced 5 percent of fed cattle. By 1983, 20 percent of fed-cattle production and processing were in the High Plains. Over that period, small cattle-feeding operations in the Corn Belt and the Central

Plains exited the industry and feedlots in the High Plains got larger. Farmer-feeders (integrated crop and livestock operations) in the U.S. declined from 45 percent of fed-cattle marketings in 1970 to 27 percent in 1981. Iowa was hardest hit, with 60 percent of farmer-feeders exiting the industry from 1970 to 1981. During this same period, substantial growth occurred in the size of individual feedlots, especially large feedlots, located predominantly in Texas, Oklahoma, and New Mexico.

Feedlots with over 16,000 head accounted for 76 percent of the total cattle fed in 1981, compared with 55 percent in 1970 (Brown); by 1994 in Texas, 57 feedlots with over 16,000 head accounted for 85 percent of the feedlot-sourced cattle marketed in the state (Texas Agricultural Statistics Service 1994). Total High Plains feedlots in 1993 numbered 124, with 111 over 5000 head. Ninety-three of these 124 feedlots were in Texas, thirteen were in Oklahoma, and five were in New Mexico (Southwest Public Service Corporation 1993).

Major factors in clustering of cattle-feeding in the High Plains were proximity to feed-grain supplies and large beef slaughter plants, favorable climate, readily available supplies of feeder cattle and proximity to southern and western markets for processed beef (Dietrich et al. 1985). Pre-1986 federal income tax laws were a factor in the construction of large corporate-owned feedlots in the High Plains (Brown 1993). In the late 1950s and early 1960s just as growth in cattle feedlots was initiated, high-yielding varieties of grain sorghum were introduced concurrent with widespread adoption of new irrigation technology, making Texas a leading producer of grain sorghum (Reimund et al. 1981). In 1966, just preceding the transition from farm-feeding of cattle to feedlots, Texas produced half of the U.S. grain sorghum crop with 80 percent being exported to other states (Holt 1966). Feedlots lowered transportation costs and increased the economic opportunities from coordinating the value-added products and services offered by allied agribusinesses.

Linkages with suppliers of stocker cattle were essential to the establishment of the feedlot industry. With little contracting or formal coordination linking them with cattle-feeding operations,[5] small cow-calf operations in Texas and the southern United States rear and raise beef cattle as stockers. According to the 1982 Census of Agriculture, small beef cattle operations in the southeast, with cow-calf farms counted all but one county in the states of Alabama, Florida, Georgia, Mississippi, North Carolina, and South Carolina (Zimet and Spreen 1986). The Southeast supplies approximately 16 percent of the stocker cattle which are fed in Texas feedlots (Dietrich et al. 1985). In 1994, 80 percent of the cow-calf operations in Texas raised less than 50 cattle, and 91 percent of Texas ranches raising stockers have less than 100 cattle. These small operators (less than 100 cattle) raised 50 percent the beef cows in Texas in 1994 (Texas Agricultural Statistics

Service 1994), and approximately 61 percent of the cattle raised in Texas feedlots were from Texas cow-calf ranches (Dietrich et al. 1985).

Cow-calf operations are labor-intensive, low-margin activities. According to standardized performance analysis of 1994 financial data from 125 Texas beef cattle herds, small operators are not likely to earn positive economic returns (McGrann et al. 1995). Tax advantages and recreational benefits are key motives for cow-calf operators in Texas. On small integrated farms in the southeast, the persistence of uneconomical cow-calf operations is best explained by its income-stabilizing role in mixed crop-livestock enterprises (Zimet and Spreen 1986). A steady supply of feeder cattle and heifers has never yet posed a constraint to growth in the cattle-feeding industry, due to the utility which small-scale ranchers derive from the cultural amenities associated with running beef cattle on small land tracts in the South and the West. Enough people like being weekend cowboys that they continue to raise beef cattle even when the enterprise is not profitable. Periodic downswings in beef cattle prices—including the 1995 drop in beef prices—do cause some marginal producers to exit, however, with re-entry by them or similar producers when prices stabilize (Bevers 1995).

Following the expansion and consolidation of feedlots in the High Plains, beef processing facilities consolidated and relocated in close proximity to new, large-scale feedlots. From 1975 to 1988, metro counties in Iowa, Illinois, Minnesota, and Wisconsin lost 14,000 meat-packing jobs, while non-metro counties in Nebraska, Kansas, and Texas gained more than 13,000 jobs (Brown 1993). Technological change was a prime mover behind locational shifts in the beef processing industry. Improved refrigeration made mechanically-processed boxed beef an alternative to traditional butcher-sourcing of packed meats (Brown 1993). More importantly, economies of size in slaughter and packing, as well as labor savings, were achieved in the new plants (Purcell 1990; Brown 1993). Ex post, a downward shift in consumer demand for beef emerges as the most significant force behind industry-wide adjustment and consolidation (Purcell 1990). Four firms controlled 51 percent of boxed-beef processing in 1979, and those same four firms controlled 79 percent in 1988. For steer and heifer slaughter, the four-firm concentration ratio increased from 37 percent in 1979 to 70 percent in 1988 (Purcell 1990). This rate and extent of consolidation is unprecedented among livestock-based agribusiness firms. A few of the same corporations involved in beef-packing and slaughter own feedlots, but more frequently they own the grain-marketing enterprises supplying feedstuffs to independent cattle-feeders.

Environmental issues

Generally speaking, cow-calf production is small in scale and low in intensity, thus any environmental side effects are widely dispersed. Because feedlots are located predominately in the semi-arid Southwest (Texas, Oklahoma, and New Mexico) non-point water pollution can be cost-effectively controlled by well-established manure management practices (Bonner et al. 1993). The beef cattle sector is not immune from complaints regarding odor and other environmental side effects, yet complaints are fewer than for other sectors due to its location in a sparsely-populated region of the country.

POULTRY AND EGG PRODUCTION

Historical trends

The organizational structure and coordination mechanisms in the poultry sector differ significantly from those used in the cattle-feeding sector. Coordination between production activities and feed distribution, as well as poultry processing, has played an important role in an industrialized broiler industry. Contracting and vertical integration was the organizational innovations employed to achieve coordination in the poultry industry, with production and marketing contracts being the most common instruments. In 1990, contract growers conducted 92 percent of broiler production, and vertically integrated firms supplied 8 percent. This transformation to a highly-coordinated poultry industry was 99 percent complete by 1970 (O'Brien 1994). From 1950 to 1990, the real price of broilers dropped from $5.28 per kilogram to $1.54 per kilogram (Westgren 1994). Within the poultry industry, the turkey sector is most concentrated with ten firms controlling 80 percent of the production. Egg production is least concentrated, with 55 leading egg producers sharing 67 percent of the market. The broiler sector is highly consolidated, with four firms accounting for 41 percent of production in 1990 (Knoeber and Thurman 1995).

Technological change and regional shifts in poultry production regions occurred in the 1950s and 1960s due to a confluence of factors (Reimund et al. 1981). Before 1950, small-scale broiler production had been a backyard enterprise in the rural-urban fringe. Male chicks from layer operations were grown out, thus production of broilers was a seasonal and erratic optional enterprise on small farms. In the late 1940s, new breeds of fast-growing chickens bred especially for the introduction of meat production which reduced feed requirements by 50 percent between 1945 and 1972. Advances in disease control (particularly vaccines against respiratory diseases) lowered death rates in larger, concentrated facilities. New engineering of broiler houses introduced mechanical cleaning and litter-removal equipment. Significant economies of size were attainable with these new engineering advances and biotechnologies. Increasingly sophisticated organization and coordination positioned poultry producers to realize their full potential, in achieving progressive cost-cutting improvements and in making efficient use of new technologies.

In 1945, 45 percent of broiler production was in the South Atlantic region, 8 percent was in the West, and the remainder was divided among the North Atlantic, the North Central, and the South Central regions, with about 18 percent each (Reimund et al. 1981). By 1971, 87 percent of U.S. broiler output was from the two southern regions. This regional shift was catalyzed by a combination of technological and organizational innovations. New southern poultry producers' competitive advantages included favorable climate and under-employed farm labor resources. Concurrent and related factors in regional shifts were new contracting opportunities with integrating firms. Production contracts at the grow-out stage in broiler production was the fundamental instrument of the broiler sector's industrialization process. Production contracts were offered by firms with linkages to both feed dealers and manufacturers as well as processing plants.

Economies of size in broiler production were achieved in both the distribution of inputs (especially feed) and in processing. Environmental regulations implemented in the 1960s made old slaughter and processing facilities obsolete (Reimund et al. 1981) and played a role in the establishment of new facilities near new clusters of producers getting started in the South. Current practices involved 150 to 300 established producers within 25 to 35 miles of an integrating firm's central structure—a hatchery, a feed mill, a processing plant and field service (Cochran 1995). In making choices about where to establish new clusters, integrating firms consider simultaneously both the prospects for recruiting and retaining grow-out contractors and the economic feasibility of a processing facility simultaneously (Howell 1995).

Production contacts in the broiler industry are risk management instruments, shifting an estimated 97 percent of the risk from growers to the integrators issuing the contract. Growers' payments depend on production outcomes rather than on market prices for their output, thus, in principle, growers bear only the "idiosyncratic" portion of production risk, estimated at 3 percent of total risk (Knoeber and Thurman 1995). Contracts with broiler companies dictate growers' choices, including genetic stock, feed regimes, and production techniques. Competence is rewarded by contract renewal. Growers bear the capital cost of investments in housing on the risk-sharing premise that the value of investment needed for processing is equivalent to the sum of grow-out investment (Westgren 1994). In effect, broiler contracts are "devices used by contractors to lease production facilities and hire labor owned by contract producers. Contractors retain title to the birds and their ownership of other production inputs is so complete as to make the contractor rather than the farmer the real producer" (Reimund et al. 1981).

Environmental issues

There are differences among contractors and across regions in whether the integrator or the grower owns the poultry litter and dead chickens. Arrangements vary from the integrator requiring specific practices and providing equipment or technical support, to contractual arrangements stating that the grower is responsible for litter and dead-bird disposal. Environmental policies and their enforcement make a difference in who takes responsibility for nutrient and waste management and its priority in the hierarchy of management tasks. Poultry litter is high in phosphorus, making it valuable as an organic fertilizer but, conversely, an ecological hazard in surface water runoff (Hays 1994). Broiler litter is often used to fertilize cropland and pastures, and is an ingredient in cattle feed. Approximately 86 percent of poultry growers in Arkansas also raise cattle (Cochran 1995). For dead bird handling, recommended practices include composting and/or storage of dead birds in freezers for transport to pet-food rendering plants. Benton and Washington counties in Arkansas produce 200 million chickens per year. Annual poultry mortality for these two counties is 3 percent or six million carcasses, thus improper disposal of dead chickens poses risks of nutrients leaching as well as bacterial contamination of groundwater (Daniel et al. 1995).

Power relationships between contractors and growers in the poultry sector have been scrutinized and indicted as an obstacle rather than a vehicle for improving nutrient management associated with poultry production. Trout fisherman are concerned that 3,700 miles of streams in Arkansas are officially listed as "fisheries threatened" and 95 percent of surface waters in the White River basin are not swimmable due to elevated coliform bacteria. Fishermen claim that "the industries involved have been totally insensitive to the [water pollution] issue" (Halterman 1993). Texas-based Pilgrim Pride, the fifth largest poultry integrator in the U.S., has been criticized for violating worker's compensation regulations and for water pollution (Cartwright 1994). Complaints against the largest poultry integrator in the U.S., Tyson Foods have ranged from water pollution to labor issues to inappropriate political contributions (Heath 1995). Tyson Foods generates an estimated 22,000 jobs in Arkansas and they control 25 percent of the U.S. poultry market. A small business owner near the firm's headquarters was quoted:

"There's been some drawbacks like water pollution and worn-out roads, but we would have dried up without them" (Heath 1995).

Though the poultry industry has been portrayed as lacking in environmental sensitivity, some recent proactive initiatives may signal potential for improvements and leadership. Tyson Foods has recently pioneered the use of centrally-located freezers for dead chickens for their contract growers in northwestern Arkansas. The dead chickens are stored under safe conditions until they can be taken to a rendering plant for pet food (Cochran 1995). The Virginia Poultry Federation has launched an initiative to improve nutrient management among growers operating in close proximity to the Chesapeake Bay (Johnson 1995).

SWINE

Historical trends

While production practices and contracting arrangements in the poultry sector have been well established and stable for over two decades, contemporary structural and regional shifts occurring in the pork sector are both rapid and dramatic. The increase in mega-farms owning over 10,000 sows has been dramatic. In 1994 there were 31 mega-farms, in 1995 there were 44 mega-farms; the top ten farms in this group added 159,500 sows in 1995, a growth rate of 19 percent (Freese 1995). Growth in large operations was matched by a decline in the number of farms by 80 percent, from over one million in 1965 to fewer than 210,000 in 1994 (Benjamin 1995). As in cattle-feeding and poultry, economies of size—as well as changes in the organization and coordination of the sector—have played important roles. Contract production of pork is increasing, from 11 percent of hog production in 1988 to 24 percent in 1994 (Grimes 1995).

Economies of size in pork production are well established (Rhodes 1995; VanArdsell and Nelson 1985), though well-managed, medium-sized integrated farms have historically achieved impressive performance as least-cost swine producers (Benjamin 1995; Strange 1988). On mega-farms, economies of size are gained from use of biotechnologies and by efficiencies gained from separating functions (with gestating and farrowing sows at one site, nursery pigs at another site, and growing/finishing hogs at a third site). Traditional pork production staged the entire process, from farrow to finish, at a single site. Separating stages in the production processes reduces the risk of diseases spreading and also allows for specialization in labor and for tailoring facility design to different functions. Technologies which require siting staged production functions in separate facilities is relatively new, thus long-standing producers (largely in the Corn Belt) are at a comparative disadvantage due to outmoded facilities which are not designed to allow full utilization of new biotechnologies. The last major influx of investment capital into the midwestern swine industry was in the 1970s, thus technological obsolescence is a problem for many independent hog producers (Hurt 1994).

Murphy Family Farms and Premium Standard Farms, among other mega-producers, coordinate the three stages of production in separate facilities located in close proximity however the facilities are owned by the same entity, and management oversight is coordinated (Warrick and Stith 1995). Another significant segment of the sector is organized quite differently with pigs being farrowed outside the Corn Belt and finished inside (Rhodes 1995), divided between independent producers and a growing sector of contracting producers. Across the U.S., hog operations are 56 percent finishing facilities, 3 percent farrow-to-finish (the traditional technology), 29 percent feeder production, and 1 percent breeding stock production, with the Iowa sector having 81 percent finishing operations and 10 percent feeder production facilities (Kleibenstein and Lawrence 1995). Contracting arrangements in the swine sector are mostly horizontal rather than vertical (Rhodes 1995). Contractors enter arrangements with growers who perform the finishing functions but these same contractors also finish hogs in their own facilities. "Hence the volume of hogs produced under contract by growers is considerably less than total contract hog marketing" (Rhodes 1995). Only an estimated 5 percent of hogs slaughtered are produced under vertically-integrated arrangements coordinating feed provision with production activities.

Relationships between economies of size in pork packing and pork production are nonetheless important factors in the location and clustering of production facilities. The average cost per pound of pork slaughter weight fell by half in the period 1975 to 1985 (Melton and Huffman 1995). Smithfield Foods of North Carolina operates the largest pork processing plant in the world, with capacity to process 8 percent of all U.S. hogs slaughtered (Hurt 1994). They coordinate production and processing in order to tailor final products to consumer preferences (Barkema 1993). Yet a coordinated relationship between production and processing is more often the exception than the rule. Cargill became involved in hog production in 1973 but did not build a pork packing plant until 1987. Many of the hogs processed by Cargill are from independent or contract growers, rather than from Cargill production facilities (Rhodes 1995). Generally speaking, it is producers getting large who then invest in packing plants, rather than packers coordinating production activities.

Hog production and processing are in the midst of structural change, thus its ultimate structure and organization are, as yet, undetermined. Pork industry observers describe current transformations and relationships between packers and producers as revolutionary, remarking that "the middle of a revolution is not a good time to predict a winner" (Looker 1995). Among the top 20 firms, ten are also involved in feed-milling, meat packing, grain merchandising, and/or transportation (Freese 1995). Presumably, these firms can reduce transaction costs and improve financial performance through coordinating functions. Contract production of hogs more than doubled between 1988 and 1994 but still accounts for only 24 percent of pork produced (Grimes 1995). Whether contracts are based on absolute performance measures or on relative performance measures makes a difference in how much risk is borne by producers, thus their propensity to accept contract arrangements in the future (Martin 1995).[6] Across regions, the prime motivation for producers who enter contracts is risk reduction rather than financial reasons (Kliebenstein and Lawrence 1995). Major sources of risk are lack of market access (Rhodes 1995) and low prices. Producers with contracts are guaranteed a buyer and a price. In the Midwest, when contract producers move in as neighbors to independent producers, guaranteed prices give contractors a competitive edge over neighbors who are independent producers, especially when market prices are low. Animosity against contract production moving into traditional swine producing regions is driven largely by fears that independent production will be crowded out (Kilman 1995).

The financial risks associated with entering a contract to finish hogs are significant. According to Kelly Zering, an Extension economist at North Carolina State University:

> "A typical hog contract farmer borrows anywhere from $200,000 to $1 million to construct his barns — a loan that's typically secured by his house and his land. But while the grower carries the debt for the 7- to 10-year life of the loan, the hog company can pull out with 30 days notice. ... While the debt is being repaid, a grower will gross about $9.50 per hog sold. After paying his debt service and expenses, the farmer is left a small amount of money for labor — roughly $7 an hour — and an additional income of about 50 cents a hog" (Warrick and Stith 1995).

In North Carolina there is a year-long waiting list for contract openings with mega-farms (Warrick and Stith 1995). Independent hog growers considering entering contract arrangements worry about the risk of their facilities becoming technically obsolete even during the period they are paying off a newly-constructed facility. The contractor could cancel their contract before investment costs are covered. Contractors are motivated by a desire to work with growers able to make best use of the newest technologies. Thus, due to rapid innovation, current contractors perceive a risk of being left with no means to recap their investment in buildings and facilities. Currently, hog contracts are designed to manage short-term production risk (Martin 1995) by guaranteeing market access, thus growers bear the long-term risk associated with capital investments in facilities.[7]

Marketing economists have various predictions about the future organization of the swine industry. Sporleder (1994), Westgren (1994), and others have speculated on the likelihood of the formation of strategic alliances—interfirm partnerships intended to improve a firm's competitive advantage over rivals—in the hog sector. Grimes (1995) reported that 18 percent of pork producers are currently involved with networking activities. Rhodes (1995) observed that earning cycles of pork packing and production cycle in opposite directions and thus often offset each other. Greater vertical integration, therefore, would allow the smoothing of cyclical earning cycles. In 1992, the general manager of Cargill's swine production department, predicted that by the year 2000 half of hogs would be produced under risk-sharing (contracting) arrangements and more producer cooperatives would be formed to take advantage of volume discounts (Mohesky 1992). He predicted 5,000 major decision makers in the swine industry by the year 2000. The shape and regional configuration of consolidation in hog production is the most rapidly-changing and the most environmentally-sensitive in animal agriculture.

Environmental issues

Environmental policies and corporate farming laws will play a role in determining on-going regional shifts and organizational changes in pork production. Iowa, still the nation's dominant hog-producing state but losing ground to North Carolina, is divided—county by county—on whether corporate farming should be allowed (Edelman 1994). Experts disagree. Breimyer (1994) predicted that mega-farms are likely to crowd out independent producers unless existing corporate farming laws are enforced and independent producers form cooperatives; Rhodes countered that if one county or state outlaws mega-farming of hogs, then large corporate producers will locate someplace where they are welcome. Pork demand is flat. Mega-farms increased supplies will drive down prices, crowding out higher-cost producers.

Environmental externalities may be the single most important determinant of future changes in the swine industry. In June 1995, 25 million gallons of hog effluent spilled into a North Carolina estuary (Smothers 1995). In July 1995 a lagoon burst and emptied 1.5 million gallons from a hog farm's lagoon into the Iowa River. "The waste killed game fish along a 40 mile stretch of river described as both scenic and pristine" (Hendricks 1995). A recent hog manure spill from a Premium Standard Farms facility into the Mussel Fork Creek in northern Missouri killed 173,000 fish and contaminated nine miles of the creek (Hendricks 1995).

Odor management associated with swine production is even more technically demanding a problem than water protection (Purvis and Outlaw 1995). Increasingly, outcries of NIMBY (not in my back yard) are the major constraint to hog facilities locating and expanding, especially in regions with dense rural populations. Odor is often the rallying cry in protests against the siting of new swine mega-farms, but the protests are driven by more fundamental issues. The most vocal opponents of large-scale corporate pork production are often themselves the heirs of multi-generational farms. They grew up raising hogs, so mega-farm investors argue that they ought to understand odor. They carry placards reading "save our family farms" and are passionate and persistent in advancing their concerns about the future of their rural communities and about corporate farms not embracing and perpetuating their values (Freese 1994). When an independent hog producer in Iowa signed a production contract, his neighbors—retired farmers themselves—expressed disdain: "He's not a farmer anymore. He's a factory" (Kilman 1995).

Though the acknowledged prime motives driving resistance against swine mega-farms are often actually the preservation of community values and family farming, forcing accountability for environmental protection is often the most effective strategy for stalling the permitting of new facilities. Odor is difficult to monitor and measure, and few states have regulations governing nuisance odor. Scientists are making progress on technical solutions (Merrill 1995a) but environmental permitting guidelines for air quality are still largely ad hoc. No federal air pollution control standards exist for CAFOs. Public hearings on the siting of new facilities often focus on water protection, focusing on the enforcement of regulations for animal agriculture which have been in place since 1972. In the absence of a venue for open debate and satisfactory resolution of the issues of genuine concern to rural residents, fears about the loss of family farms may be driving insistence on increasingly stringent standards for water protection or nuisance odor control for any new or expanding animal production facilities. Iowa State University's dean of the College of Agriculture summed up the issues facing communities seeking a balance between environmental stewardship and the economic sustainability of rural lifestyles and family farming: "I'm worried about towns getting torn apart" (Kilman 1995).

Ironically, it is the largest hog production facilities who have a comparative advantage in environmental compliance. Cutting-edge technologies for treating water and controlling odor involve significant economies of size (Purvis and Outlaw 1995; Van Horn 1995). Moreover, large-scale producers are likely to be affiliated with others who have expanded recently and built new facilities. Relocation is an opportunity for experimentation and learning with new technologies. Technological efficiency sufficient to satisfy neighbors continues to elude even the newest facilities in the Midwest. Most recently, new expansions are re-locating to sparsely-populated sites hungry for economic development. The newest mega-farm, Circle Four in southeastern Utah, has placed a three-mile set-aside between its facilities and its nearest neighbors (Marberry 1994). Premium Standard Farms is planning an expansion of their Dalhart, Texas, facility—18 miles outside town and the opposite direction of the prevailing winds. The current trend in environmental policies pertaining to animal agriculture is federal guidelines implemented by states and enforced, by necessity, at the local level. Without institutional innovation or well-orchestrated interventions in community-level decision processes, mega-farms are likely to continue to relocate in sparsely-populated, arid regions. This trend has major implications for future change in the swine industry's regional location patterns. The current propensity toward hog production moving out of the Midwest and into sparsely-populated regions with fewer protests and conflicts has potentially sizable long-run economic, sociological, and ecological ramifications.

DAIRY

Historical trends

In the past decade, trends in dairying are toward fewer and larger dairy farms producing more milk—per cow and per farm—using technologies which exploit economies of size. Average annual efficiency increases in dairying have been 2 to 3 percent, owing largely to technological improvements such as improved genetics and specialized labor (OTA 1991). Regional shifts have been out of traditional dairy regions—the Great Lake States and the Northeast—to warmer, more arid locations in the Southwest and the West (Fallert et al. 1994). California became the top dairy producing state in 1993, outpacing Wisconsin. Between 1982 and 1992, the fastest growing dairy state was New Mexico, whose herd increased by 62 percent.

Since the 1920s the production and marketing of dairy products has been coordinated through a federal milk market order (Babb et al. 1983). Vertical integration, therefore, has played a limited role in recent regional shifts, though proposed market-oriented reforms proposed under the 1996 farm bill provisions are likely to open the dairy industry to locational shifts driven by relative cost-effectiveness across regions. Cost-effectiveness in dairying is best achieved when milk is produced near large population centers and near processing plants in order to save on transportation costs. The purpose of the federal milk marketing orders has been to facilitate such coordination.

The major impetus behind recent regional shifts in dairying has been the adoption of new technologies achieving cost efficiencies through economies of size and through labor specialization. In California and the Southwest, the prevailing dairy systems specialize in milk production and import grains and forages grown by others, sometimes in other regions. These producers specialize in milk production and channel their management expertise into animal health and nutrition. When several large dairies are located cluster in close proximity, this cluster becomes a market for a specialized infrastructure (Pagano and Abdalla 1994)—including veterinarians, dairy nutritionists, dairy equipment suppliers and repair services, and accountants—services essential to cost-effective large-scale dairy production.

Erath County, Texas, is a classic example of a cluster. Large-scale dairy producers in central Texas emigrated from California in large numbers in the 1980s. Estimated growth in the dairy herd from 1980 to 1990 was 73 percent; currently, there are approximately 70,000 cows within a 50-mile radius (Gerlin 1994). Texas dairy producers resent the sweeping generalization that they fled California's stringent environmental laws, seeking a more laissez-faire situation in the Texas of the early 1980s. They moved to Texas seeking inexpensive land, low-cost feedstuffs, and proximity both to a dairy service infrastructure (such as veterinarians and nutritionists) and to urban amenities (the Dallas-Fort Worth metroplex). In the early 1990s, many California dairy emigrants avoided Texas and instead located new facilities in New Mexico, Idaho, Utah, or Kansas. Today's Texas dairy leaders are adamant that successful dairying does not require a lax regulatory posture regarding environmental compliance. Rather, they have urged Texas legislators to work toward a long-run vision of a sound business environment with a judiciously-enforced and technically-sound set of environmental regulations. Individual dairy producers have also learned the importance of their making personalized efforts to accommodate their neighbors' and community's preferences (Stalcup 1994).

As in swine production, due to economies of size and availability of capital, large-scale dairy producers often enjoy a comparative advantage in environmental compliance over smaller producers operating in older dairy facilities (GAO 1995). However, for large-scale dairy producers, because their capital costs are high and investments in a particular technology, once made, are irreversible, there are benefits from a regulatory environment which is predictable, technically-sound, and pragmatic (Pagano 1993; Pagano et al. 1994).

Environmental issues

As is the case in the swine industry, the nature and enforcement of environmental compliance policies is likely to play a key role in future regional shifts in the dairy industry which are being driven by a confluence of economic and policy forces as well as by environmental concerns. Local-level protests about the environmental side-effects of clustering have influenced the environmental compliance requirements facing large-scale dairies and other CAFOs. In 1992 in Erath County, Texas, spring rainfall was well above average and though there was never a 24-hour, 25-year flood event, several consecutive days of rainfall resulted in 34 dairy lagoons spilling into adjacent streams and rivers. The Sierra Club threatened to file a citizen's lawsuit against the discharging dairies. Instead, the EPA's Region VI issued a special permit which scaled up pollution prevention requirements. Beginning in 1994, Texas dairy producers have been required to comply with three separate permits: state air and water permits, plus the EPA special permit. In Texas, any sitings of new facilities or expansions of existing facilities requires a series of public hearings until neighbors and regulators are convinced that the technical requirements are satisfied. In December, 1993, a dairy producer willing to meet and exceed (more than double) all the technical requirements specified by all three required permits was denied the opportunity to build a new 2020-cow dairy due to opposition from his neighbors (Holan 1993). The only Texas producer siting a new 400-cow dairy in 1992 negotiated with neighbors in a dozen public hearings, spending over $100,000 in legal fees and two years in the process (Stalcup 1992). This high-pitched local-level conflict is not unique to Texas: for example, in rural north-central Florida, over 2,000 neighbors signed a petition against the siting of a new 750-cow dairy in Dixie County (Ritchie 1995).

Perhaps the most significant recent change in policies pertaining to large-scale animal agriculture resulted from recent judiciary action in upstate New York, *Southview Farms v. concerned area residents for the environment*. The opinion issued by the U.S. District Court of Appeals (the New York district) ruled that a 2200-cow dairy farm was a point source of pollution, **including** the adjacent fields where forage crops are raised using manure as an organic fertilizer (Merrill 1995b). The rural residents who sued their neighbor filed the lawsuit on grounds of water pollution. Originally, a local jury decided the case in favor of the dairy farmer—they were not convinced that dairy effluent was responsible for alleged groundwater and surface water pollution—but in an appeal that ruling was overturned (Merrill 1993). The New York Commissioner of Agriculture and Markets observed that "this case was really about odor" (Merrill 1995b). The dairy farmer and the New York farm organizations supporting him spent $600,000 in legal fees and further appeals are planned (Roenfedt 1995).

The Southview case is significant because it signals an abrupt shift in thinking about the scope and boundaries of large-scale animal agriculture's environmental protection responsibilities. Since the 1972 Clean Water Act was passed, large-scale confined animal feeding operations (CAFO) have been required to handle surface water runoff under a zero-discharge permit modeled after the point-source requirements enforced for industrial producers. The Southview ruling extends the pollution prevention responsibilities of dairy producers and other CAFO operators significantly. Spreading manure on cropland is now also under the purview of the NPDES permit. This legal precedent makes farmers accountable and liable for whole-farm nutrient management. Technically this change in the NPDES permit would require a producer to be accountable for the fate and transport of all the phosphorous and nitrogen flows associated with manure management. If enforced to the letter of the law, new environmental monitoring expenses for both producers and regulators would be exorbitant. In the meantime, until further litigation and rule-making unfold, communities can demand significantly more of new and existing CAFOs in public hearings.

THE ENVIRONMENTAL EFFECTS OF CLUSTERING

To summarize, a vivid pattern is emerging of bigger and fewer animal agriculture production facilities. The primary impetus for clustering is economic coordination, economies of size in production and processing technologies, and savings in transportation costs. From a community development standpoint, economic activity from animal agriculture expansion is a welcomed mechanism for recruiting investment capital and employment opportunities to rural areas. Shrinking rural communities compete to attract expanding animal agriculture production facilities and allied agribusinesses, and states progressively losing producers or processors strategize to keep or revitalize their industries (Freese and Fee 1994). The dislocation associated with these industrialization processes are sources of community conflict.

State and federal environmental compliance requirements have been ratcheted up due to pressure resulting from community-level protests which lead to litigation or threatened lawsuits. The pollution prevention requirements in place when animal agriculture arrives in a community often become insufficient because the environmental effects of clustering are cumulative rather than immediate (Pagano and Abdalla 1994). The manure management arrangements of a single, established facility may be environmentally sound and acceptable when permitted, but as that facility grows to take advantage of economies of size in new technologies and as new entrants compete for cropland, airsheds, and watersheds, the assimilative capacity of the community becomes over-taxed. Localized nutrient overloads are most serious when feedstuffs are imported rather than manure being used as an inorganic fertilizer for raising forages or grains to feed livestock (Lanyon 1995). Existing facilities are often grandfathered in, which increases the difficulties faced by investors trying to site new facilities with improved pollution prevention technologies. Moreover, the political dynamics associated with public hearings work against experimentation with new technologies (Pagano et al. 1994). Coordination mechanisms (federal, state, and local policy design and enforcement as well as collaboration with industry groups and allied agribusinesses) are urgently needed to improve what is currently an ad hoc and capricious set of environmental compliance policies governing animal agriculture. Without concerted efforts to coordinate, there will be a continuing regional shifts in animal agriculture with "difficult policy trade offs between clean air and economic development being made rather hastily on a state and/or local level" (Rhodes 1995).

Christ (1995) capsulated the problem of striking an appropriate balance among tradeoffs as involving four competing sets of expectations for animal agriculture: (1) least-cost production, (2) economic growth and development, (3) environmental protection, and (4) minimization of economic dislocation. Different states—even different localities—are currently choosing to balance these priorities differently. Neither communities nor regulators nor producers nor allied agribusinesses nor vertically integrated firms involved with animal agriculture are satisfied with the current quagmire of environmental compliance where the battlegrounds are often staged over environmental compliance issues but the war is about more complex issues involving industrialization and rural/farm policies.

TOWARD A RESEARCH AGENDA

For large-scale producers, vertically integrated firms, and allied agribusinesses the problems with environmental compliance would not be solved immediately and automatically if the current patchwork of state and federal regulations were replaced by a coherent set of rules on both conflict management processes and performance standards. The cumulative nature of environmental externalities from animal agriculture—a self-reinforcing phenomenon—further exacerbates the difficulty of crafting a stable and sustainable future set of policies.

This chapter has framed the problem of satisfying environmental stewardship responsibilities as a perpetual moving target due to changing relationships with communities and unstable preferences regarding the requirements for an acceptable industrialized animal agriculture, as well as increasing knowledge about technology options and the overall ecological ramifications of clustering. This overview of animal agriculture suggests that significant differences exist across sectors. Moreover, not enough is known about existing institutional dynamics nor about firm-level decision-making to fully consider policy issues and options involving privatization of information and technology. In addition to better data, tackling the policy research challenge will require alternative models and analytical tools to aid in the formulation and modification of an evolutionary yet coordinated set of institutions and policies, responsive to high-quality technical, ecological, economic, and sociological complexities and interdependencies. These issues are inherently cross-disciplinary.

Three themes for future inquiry are proposed as an integral part of a larger research agenda on privatization and information issues associated with the industrialization of agriculture.

Policy linkages

An emergent theme from the recent history of animal agriculture is fragmentation in policy design and enforcement. The task of managing the environmental effects of industrialization in the livestock and poultry sectors is currently spurring a proliferation of diverse and uncoordinated firm-level responses conditioned by interactions with local-level institutions (both government and allied agribusinesses). Ostrom (1995) presented a persuasive case that the greater the complexity associated with an environmental management situation, the greater the sophistication required in overall institutional design. From her comparative analysis of the successful design for long-sustained and self-governing resource management systems, she distilled a set of desirable properties and principles for grassroots-based institutional design. "When resource appropriators design their own operational rules to be enforced by individuals who are local appropriators or accountable to them, using graduated sanctions that define who has rights to withdraw from the resource and that effectively restrict appropriation activities given local conditions, the commitment and monitoring problems are solved in an interrelated manner" (Ostrom 1995).

Though local empowerment has proven most effective in coordinating decisions about managing complex and interdependent natural resource allocation issues with the other goals and values of communities, Ostrom warned that the policy prescription is not as simple as dictating a trickle-up philosophy. For federalism to work, as the complexity of environmental and economic interdependencies increases, web-like institutional design is required. Specifically, linkages to higher levels of government and agencies with strong technical capacity are essential. "If all local communities were to have to develop their own scientific information about the physical settings in which they were located, few would have the resources to accomplish this" (Ostrom 1995). As the complexity of environmental effects and the diversity of technological options gets greater, the demand increases for institutional resiliency and pragmatism and, perhaps most important, synchronization of the roles and functions performed at the firm, local, state, and federal levels.

Porter and van der Linde (1995) proposed general guidelines for design and implementation of environmental regulations. They recommended "phasing environmental rules as goals that can be met in flexible ways, encouraging innovation to reach and exceed those goals, and administering the system in a coordinated way" (p. 110). Boggess and Cochran (1995) analyzed the particular challenges associated with formulating an evolutionary framework to monitor and mitigate the environmental effects from animal agriculture. Smith (1995) and Batie (1995) discussed institutional design issues for animal agriculture, both arguing persuasively that states are taking leadership in defining and modifying the institutional opportunity sets within which animal agriculture operates. Taken together, these policy analysts have formulated the elements of a robust conceptual framework for studying and advising states and localities who are crafting and carrying out environmental compliance policies governing animal agriculture. Adaptive management is likely to involve incremental modifications of policies. To facilitate pragmatic and intelligent tinkering, the logical next step is nitty-gritty empirical analysis of real-world policy experiments drawing and building on methodologies proposed by Campbell (1977), Schmid (1987), and Lee (1993). Design of mechanisms to archive and apply lessons learned is part of adaptive management.

Understanding and modeling economic behavior under complexity

The fundamental determinant of policy effectiveness is the responsiveness of individual producers and allied agribusiness partners to policy signals. A second proposed theme for research on industrialization is empirically-based inquiry on decision making about environmental compliance, and about negotiations with neighbors and regulators. For such research to generate results which are useful to producers and allied agribusinesses, as well as to agencies and analysts concerned with effective design and implementation of environmental regulations, detailed and frank interviews with front-line decision makers will be essential. In the recent past, data constraints have hobbled empirical research on micro-level economic behavior in consolidating industries. This phenomenon is easy to explain: closely-held information is a source of strategic advantage in competitive industries. Consolidation in animal agriculture has been a rapid and competitive process. Accordingly, this competition has worked against the open sharing of information about intra-firm and inter-firm decision making. Marketing economists interested in agribusiness industrial organization have a sophisticated understanding of the incentives of large firms in growing industries to hoard information about their internal mechanics (Babb and Lang 1983; Boehlje 1995). Robust theories exist for analyzing intra-firm and inter-firm relationships, the most recent and comprehensive compendium being Milgrom and Roberts (1992). Despite these constraints, Extension farm management economists have accumulated significant knowledge about the inner workings of animal agriculture, in tandem with their efforts to adapt their messages and information delivery to accommodate the needs of executive managers (McGrann 1993; Holt 1971).

Regarding behavior in response to environmental regulations, little empirical data is available to supplement the anecdotal stories documented in the farm press. Only recently have the local-level situations facing animal agriculture become sufficiently pressurized and exasperating that there may be a new openness to discussing decision-making about environmental compliance with researchers. Such openness might make producers and allied agribusinesses willing to share the data required to test these hypotheses. Conflicts involving environmental compliance are becoming an intractable obstacle to expansion and economic progress for firms on the leading edge of industrialization in animal agriculture. Perhaps exasperation will lead to common ground for creative problem-solving about how existing contract arrangements and information flows might be used to improve animal agriculture's future ability to achieve sustainable and acceptable environmental stewardship in tandem with their other economic goals and investments.

Producers and associated allied agribusinesses would need to see the potential benefits of such research in order to agree to collaborate. They will need to be convinced about their participation having a direct impact on policy design and enforcement. An iterative and pragmatic research design with deliberate linkages to policy innovation will be essential. The most challenging aspect of initiating and carrying out research involving intra-firm and inter-firm coordination to achieve common goals with a long-term orientation toward community-driven sustainability.

As a starting place, empirical testing is recommended to verify or refute four working hypotheses which have been proposed about the behavior of industrializing animal agriculture in response to tightening compliance requirements. First, empirical research is needed to explore how proximity of industrializing animal agriculture production facilities might be viewed by neighbors and regulators as friend rather than as foe. As the size of animal agriculture production facilities gets larger, anecdotal evidence would suggest that relationships with neighbors and regulators seem more likely to be antagonistic rather than collaborative. Local-level volatility and idiosyncratic conflicts become major obstacles to installing the most efficient and cost-effective technologies. Thus it is where current circumstances are the most polarized that incentives on all sides are the greatest to work together to improve upon the site-specific context for achieving effective environmental stewardship and protection. It is a testable hypothesis that intra-firm communication channels within vertically integrated firms as well as emerging networking arrangements among independent producers located in clusters are potential avenues for organizing collaborative efforts to achieve effective management of nutrients, odor, flies, and other environmental side effects from animal agriculture.

Second, opportunities exist to learn from earlier experiences with the environmental regulation of factories, where the character of industrialization has exhibited similarities and differences compared with agriculture. Cartelization occurred when point-source polluters, such as hydroelectric power plants and paper mills, were asked to internalize the costs of controlling their emissions. Policy guidelines for environmental technologies were tested and fine-tuned by the largest and most powerful firms. Their compliance technologies were codified into technology-based rules, thus their chosen compliance measures crowded out alternative approaches. It is a testable hypothesis that large-scale livestock or poultry producers might similarly establish and influence policies biased in their favor (Purvis and Outlaw 1995), thus accelerating consolidation in animal agriculture and diminishing opportunities for experimentation with alternative technological options.

Third, not only the content but also the clarity of policy signals makes a difference in the likely behavior of the regulated community. Evolving policies and enforcement protocols were a key source of uncertainty which delayed Texas and Florida dairy producers' investments in innovative environmental compliance technologies (Purvis et al. 1995). It is a testable hypothesis that indeterminateness in environmental regulations and their implementation has been an obstacle to investment and experimentation with new technologies in other livestock and poultry sectors, as well as in other regions.

Finally, it is a testable hypothesis that large-scale and/or vertically-integrated producers are less likely to share new technological innovations than would smaller producers—those who have historically operated farms with integrated crop/livestock enterprises—served by public technology-transfer agencies such as the Cooperative Extension Service and the Natural Resources Conservation Service (Smith 1995). This premise could be refuted (or confirmed) by evidence of technology sharing (or the contrary) on the part of the producers and firms leading growth in animal agriculture. The degree to which technology development for environmental compliance has been a site-by-site—thereby idiosyncratic—process has implications for the public-private balance in information development and delivery for animal agriculture in the future. An empirically-based analysis of decision processes on technology selection and adaptation over the past fifteen years would aid in current decisions about how to allocate technical assistance efforts targeting in animal agriculture by Extension and the Natural Resource Conservation Service.

The public-private balance in information delivery

The final research theme recommended is a research initiative to collect a precise specification and priority-ranking regarding how technical experts employed in the public sector—in particular, Extension and the Natural Resources Conservation Service—can best allocate limited resources (especially human capital) to provide needed services and information which the private sector is unlikely or unable to offer.

Perry (1995), Lippke and Rister (1992), Harris and co-authors (1992) and Schnitkey and co-authors (1992) have documented that indeed large numbers of large-scale producers have contact with Extension and value their assistance. While this body of survey research is valuable, it provided no detailed data about the specific issues associated with environmental stewardship being addressed nor about how such information or its delivery could be improved. Further survey research—building on these earlier studies—is needed to collect more detailed information about the nature of the interactions between industrializing producers and public-sector information providers (Extension and the Natural Resource Conservation Service) as well as about redundant services, emerging issues, and unmet needs. The USDA has recently embraced the challenge of involving and servicing a broader client groups involving groups of consumers, community leaders, neighbors of livestock producers, representatives of environmental groups, and urban taxpayers. Moreover, scoping exercises to assess and respond to the needs of new clientele—in particular, non-farm rural residents, concerned urban residents, and environmental activists—as well as the new needs of community leaders, merit consideration and high priority in shaping the future agendas of Extension and the Natural Resources Conservation Service.

A promising and emerging area of activity for innovative Extension specialists and county agents is in brokering information between environmental regulators and animal agriculture producers and in supplying public issues education to other stakeholders including non-farmers. Roy Carriker, an Extension economist at the University of Florida in Gainesville, is leading a design team to present a series of training sessions for county agents on environmental regulations. Deanne Morse, the dairy Extension agent at the University of California in Davis, considers environmental regulators her key client group. In Texas and Pennsylvania,

Extension engineers and agronomists—John Sweeten and Bill Harris, and Bob Graves and Doug Beegle, respectively—have played leadership roles in the iterative and on-going process of designing BMPs and nutrient management planning procedures for livestock and poultry producers. A multi-disciplinary group of social scientists and entomologists and engineers at Penn State University mobilized a permanent team to work with communities in conflict over fly outbreaks at poultry operations in the summertime (Purvis and Abdalla 1995).

Perhaps the most innovative Extension leadership has emerged at the county level, where agents have been pragmatists in the forging of public-private partnerships for collaborative problem-solving at the community level where clusters are developing. Noteworthy examples include Joe Pope, the county agent in Erath County, Texas; Judith Wright, the county agent in Cayuga County, New York (Wright 1995), and Bill Thompson, the county agent in Chaves County, New Mexico, among dozens of others across the U.S. If trends toward federalism continue and thus the role of grassroots institution-building grows, then systematic analysis of early efforts at information brokerage for purposes of assisting with environmental compliance is an opportunity to learn from successes and failure. In analyzing these experiments in institutional innovation and charismatic adaptive management, the leadership roles of individuals are a prominent feature. Key policy research issues concern institutional design: Are organizations and agencies doing the best possible job of supporting the innovative leadership of individuals? Are there cases where individuals struggle against institutional constraints? Where constraints exist, they are often the product of history, thereby intertwined with idiosyncratic local and larger circumstances.

CONCLUSION

The historical account of industrialization in animal agriculture developed in this chapter was based largely on retrospective empirical analysis and on policy accounts reported in the farm press. A theme which runs through the recommendations for policy research is the need for better data about dynamically-changing institutions and firm-level decision-making. Particularly in the environmental compliance arena, policy decisions at the local, state and federal level could benefit from an infusion of real-world evidence and implications, as outlined above. Though environmental management is not the only challenge facing front-line decision makers in animal agriculture, it seems a likely starting place for engaging cooperation and collaboration with researchers, as well as the agencies and institutions charged with managing the environmental effects of an industrializing animal agriculture.

NOTES

[1] Horizontal relationships are linkages among enterprises performing similar functions (such as production, processing or marketing). Vertical relationships are linkages among enterprises performing different functions (such as the same manager or firm coordinating both production and processing activities). Horizontal or vertical integration describes a continuum of institutional arrangements involving a single firm, a cluster of firms, or some organization which coordinates multiple operations performing similar or multiple functions, respectively.

[2] An open question is whether facilities achieving satisfactory or exemplary environmental management are desirable or acceptable as neighbors. Anecdotal evidence would indicate a bias in some communities against all expansion of large-scale agriculture, regardless of its success in minimizing environmental side effects.

[3] In a recent survey article, Jaffee and coauthors (1995) examined the "widespread belief that environmental regulations have a significant effect on the siting of new plants in the United States" (p. 149). The limited empirical work has been conducted on industrial plant location in the United States would indicate that "these concerns may not be well-founded" (p. 149). Disaggregated data needed to analyze these issues systematically are a constraint to this line of empirical inquiry, in the US manufacturing sector as well as in agriculture.

[4] All concentrated animal feeding operations (CAFOs) — including dairies milking over 700 cows, swine farms with over 2500 sows, and beef feedlots with over 1000 cattle — are required to hold a National Pollution Discharge Elimination permit. The requirements for compliance are satisfied by CAFOs with sufficient wastewater storage capacity to accommodate a 24-hour, 25-year rainfall event. The permit stipulates no allowable discharges of wastewater (including runoff from a rainfall event) from a CAFO into the waters of the United States. Compliance with an NPDES permit generally means constructing an anaerobic lagoon. Enforcement of NPDES permits, has been the responsibility of the state-level environmental regulatory agency. Implementation procedures for NPDES permits and the guidelines for applying manure to cropland vary considerably across states (Outlaw et al. 1993). By 1995 estimates, 1,987 of an estimated 6,600 CAFOs in the United States hold federally-administered NPDES permits (GAO 1995). The remainder are either permitted by state environmental regulatory authorities or do not hold permits.

[5] As of 1990, 12 percent of fed cattle were produced under contract and 4 percent were produced in production facilities owned and managed by vertically-integrated firms (O'Brien 1994).

[6] Absolute performance measures in marketing contracts are pre-agreed weight gains or production results. Relative performance measures in marketing contracts are based on comparisons with the production results of other producers.

[7] In the Midwest, historically returns to investments from pork production have been favorable thus this risk-bearing arrangement worked in favor of pork producers. In Iowa since 1980, estimated average annual returns from capital investments in the pork industry were above 25 percent (Hurt 1994). In Missouri, annual returns from pork production averaged 13.5 percent for 1983 to 1992 (Breimeyer 1994). With the abrupt down-turn in pork prices in the mid-1990s, however, independent pork producers suffered significant losses and risk of technical obsolescence seemed daunting.

REFERENCES

Abdalla, C., L. Lanyon, and M. Hallberg. 1995. What do we know about historical trends in firm location decisions and regional shifts: Policy issues for an industrializing animal agriculture. American Journal of Agricultural Economics. Volume 77:5 (December).

Babb, E., R.D. Boynton, W. Dobson, and A. Novakovic. 1983. Milk marketing orders. In Federal Marketing Programs in Agriculture: Issues and Options, W. Armbruster, D. Henderson, and R. Knutson (eds.). Danville, IL: Interstate Printers & Publishers, Inc.

Babb, E., and M. Lang. 1983. Intrafirm decision making: Private and public consequences. In Future Frontiers in Agricultural Marketing Research, Paul L. Farris (ed.). Ames, Iowa: Iowa State University Press.

Barkema, A. 1993. Reaching consumers in the twenty-first century: The short way around the barn. American Journal of Agricultural Economics. 75:5 (December): 1126-1131.

Batie, S. 1995. Future environmental policy: The state role. Presented at a Symposium on the Influence of Social Trends on Agricultural Natural Resources. Washington, D.C., May 21-June 2.

Benjamin, G. 1995. The changing hog sector. AgLetter. Federal Reserve Bank of Chicago, April.

Bevers, S. 1995. Ridin' out the storm. The Cattleman. July.

Boehlje, M. 1995. Discussion: Vertical coordination and structural change in the pork industry. American Journal of Agricultural Economics. Volume 77:5 (December).

Boggess, W., and M. Cochran. 1995. Multiple policy instruments: An evolutionary approach to animal waste management. In Animal Waste and the Land-Water Interface, Kenneth Steele (ed.). New York: Lewis Publishers.

Bonner, B., W. Harman, and S. Amosson. 1993. Texas High Plains feedlots: Survey of 1992 characteristics and manure management practices. Texas Agricultural Experiment Station. Bulletin PR-5103. Texas A&M University, November.

Breimyer, H. 1994. Do mega hog farms foretell farming of the future?: Observations from a policy forum. In Economic and Policy Information for Missouri Agriculture: A Newsletter. Columbia, Missouri: University of Missouri, Department of Agricultural Economics. Volume 38, Number 5, September/October.

Brown, D. 1993. Changes in the red meat and poultry industries: Their effect on nonmetro employment. Washington, DC: USDA, Economic Research Service, Agricultural Economic Report Number 665.

Campbell, D. 1977. Reforms as experiments. In Readings in Evaluation Research. 2nd edition. Francis G. Caro (ed.). New York: Russell Sage Publishers.

Cartwright, G. 1994. The baron of Texas agriculture: Bo Pilgrim. Texas Monthly. September.

Christ, P. 1995. Policy issues for an industrializing animal agriculture: Comment. American Journal of Agricultural Economics. Volume 77:5(December).

Cochran, M. 1995. Animal production and sustainable agriculture: The poultry sector. Remarks presented at an organized symposium, American Agricultural Economics Association meeting, Indianapolis, Indiana.

Daniel, T., D. Edwards, D. Nichols, K. Steele, and S. Wilkes. 1995. Water quality and poultry disposal pits. Fayetteville, Arkansas: Arkansas Water Resource Center, Fact Sheet Number 2.

Dietrich, R., P. Thomas, and D. Farris. 1985. The Texas cattle feeding industry—Operations, management and costs. College Station, TX: Texas Agricultural Experiment Station, Bulletin B-1495.

Edelman, M. 1994. State uniformity versus local flexibility in hog confinement. Plain Common Sense: Iowa State University Extension News. 31 October.

Fallert, R., M. Weimar, and T. Crawford. 1994. Here's why milk's moving west. Hoard's Dairyman. 10 January.

Freese, B. 1994. Fed up with the big boys. Successful Farming. April.

Freese, B. 1995. What a difference a year makes: Pork powerhouses 1995. Successful Farming. October.

Freese, B and R. Fee. 1994. Livestock hungry states: Some states offer carrots, others turn their backs on livestock expansion within their borders. Successful Farming. January.

General Accounting Office. 1995. Animal agriculture: Information on waste management and water quality issues. Briefing Report to the Committee on Agriculture, Nutrition, and Forestry, US Senate. GAO/RCED-95-200BR.

Gerlin, A. 1994. As the state's milk production increases, so do the screams of environmentalists. Wall Street Journal. 25 May.

Grimes, G. 1995. Under the microscope. Check-off. (National Pork Producers Council magazine) November/December.

Halterman, S. 1993. The politics of Arkansas. Trout. Winter.

Harris, H., M. Hammig, J. Jordan, D. Smith, and E. Kaiser. 1992. Use and perceptions of Extension programs by farmers in four Southern states. Extension Bulletin #EER-136, Clemson University, February.

Hays, S. 1994. A cleanup for poultry litter. Agricultural Research. May.

Heath, T. 1995. When the foxes start sniffing around the henhouse: Tyson Foods has found that national policies and poultry don't mix. Washington Post Weekly. 31 July.

Hendricks, M. 1995. Manure spills threaten waterways. Kansas City (Missouri) Star. 24 September.

Holan, M. 1993. Tuls dairy permit denied: Third denial this year comes despite environmental plans. Stephenville (Texas) Empire-Tribune. 17 December.

Holt, J. 1966. An economic analysis of confinement pork producing systems on the South Plains of Texas. M.S. thesis. Texas Tech University, May.

Holt, J. 1971. Agriculture gets bigger—So what? Presentation to the Southern Extension Farm Management Committee, 14 April.

Howell, D. 1995. Fowl play: A bird in the hand? Bryan (Texas) Eagle. 27 April.

Hurt, C. 1994. Industrialization in the pork industry. Choices. Fourth Quarter.

Jaffee, A., S. Peterson, P. Portney, and R. Stavins. 1995. Environmental regulation and the competitiveness of U.S. manufacturing: What does the evidence tell us? Journal of Economic Literature. XXXIII: 132-163.

Johnson, J. 1995. Why Virginia poultry industry is committed to nutrient management. Bay Journal. June.

Kilman, S. 1994. Iowans can handle pig smells, but this is something else: Giant hog 'factories' strain inherent neighborliness of a rural community. Wall Street Journal. 4 May.

Kliebenstein, J. and J. Lawrence. 1995. Contracting and vertical integration in the U.S. pork industry. Staff Paper #265, Iowa State University.

Knoeber, C. and W. Thurman. 1995. Don't count your chickens ...': Risk and risk shifting in the broiler industry. American Journal of Agricultural Economics. 77:3: 486-496.

Lanyon, L. 1985. Does nitrogen cycle? Changes in the spatial dynamics of nitrogen with industrial nitrogen fixation. Journal of Production Agriculture. 8: 70-78.

Lee, K. 1993. Compass and Gyroscope: Integrating Science and Politics for the Environment. Washington, DC: Island Press.

Looker, D. 1995. From the powerhouse to the packinghouse: Strong ties bind giant pork producers and packers. Successful Farming. March.

Marberry, S. 1994. Hog industry insider. Feedstuffs. 7 November.

Martin, L. 1995. Pork ... the other white meat? Production contracts, risk shifting, and relative performance payments in the pork industry. Michigan State University, East Lansing, Agricultural Economics Staff Paper, #95-48.

McGrann, J. 1993. Fee based Extension programming in the southern region. Presented to the Southern Regional Farm Management and Marketing Committee. San Antonio, Texas, June,.

McGrann, J., S. Bevers, L. Falconer, R. Gill, and J. Parker. 1995. Cow-calf producers in West Texas are more competitive than East Texas. Texas Agricultural Extension Service, IRM-SPA Handbook. 21 July.

Melton, B., and W. Huffman. 1995. Beef and pork packing costs and input demands: Effects of unionization and technology. American Journal of Agricultural Economics. 77,3: 471-485.

Merrill, L. 1995a. Odor and fly concerns will challenge us. Hoard's Dairyman. July.

Merrill, L. 1995b. New York manure case reversal raises regulatory concerns. Hoard's Dairyman. 10 January.

Merrill, L. 1993. No clear winners in Southview farm manure suit. Hoard's Dairyman. 25 August.

Milgrom, P., and J. Roberts. 1992. Economics, Organization, and Management. Englewood Cliffs, NJ: Prentice Hall.

Mohesky, R.H. 1992. Here is Cargill's vision of the pork industry by 2000. Successful Farming. April.

O'Brien, P. 1994. Implications for public policy. In Food and Agricultural Markets: The Quiet Revolution, L.P. Schertz and L.M. Daft (eds.). Washington, DC: National Planning Association.

Office of Technology Assessment, U.S. Congress. 1991. U.S. dairy at a crossroad: Biotechnology and policy choices—Special report. OTA-F-470. Washington, D.C.: U.S. Printing Office.

Ostrom, E. 1995. Designing complexity to govern complexity. In Property Rights And The Environment: Social And Ecological Issues, S. Hanna and M. Muasinghe (eds.). Washington, D. C.: World Bank.

Outlaw, J., R. Schwart, Jr., R. Knutson, A. Pagano, A. Gray, and J. Miller. 1993. Impacts of dairy waste management regulations. Agricultural Food Policy Center Working Paper, Number 93-4, Texas A & M University.

Pagano, A. 1993. Ex ante forecasting of uncertain and irreversible dairy investments: Implications for environmental compliance. Ph.D. dissertation, University of Florida.

Pagano, A., and C. Abdalla. 1994. Clustering in animal agriculture: Economic trends and policy. In Balancing Animal Production and the Environment. Proceedings of the Great Plains Animal Agriculture Task Force. Denver, CO: October 19-21.

Pagano, A., K. Sims, J. Holt, W. Boggess, and C. Moss. 1994. Environmental permitting and technological innovation. In Balancing Animal Production and the Environment. Proceedings of the Great Plains Animal Agriculture Task Force. Denver, CO: October 19-21.

Palmquist, R., F. Roka, and T. Vukina. 1995. Hog operation, environment effects, and residential property values. Unpublished manuscript, North Carolina State University, Department of Agricultural Economics.

Perry, J. 1995. Farmers' business contacts with USDA agencies. ERS Briefing Paper. Washington, DC: USDA, Economic Research Service.

Porter, M., and C. van der Linde. 1995. Toward a new conception of the environment-competitiveness relationship. Journal of Economic Perspectives. 9:4: 97-118.

Purcell, W. 1990. Economics of consolidation in the beef sector: Research challenges. American Journal of Agricultural Economics. 72,5: 1210-1218.

Purvis, A., and C. Abdalla. 1995. Analyzing manure management policy: Toward improved communication and cross-disciplinary research. In Animal Waste and the Land-Water Interface, K. Steele (ed.). New York: Lewis Publishers.

Purvis, A., W. Boggess, C. Moss, and J. Holt. 1995. Technology adoption decisions under irreversibility and uncertainty: An ex ante approach. American Journal of Agricultural Economics. 77,3: 243-250.

Purvis, A., and J. Outlaw. 1995. What we know about technological innovation to achieve environmental compliance: Policy issues for an industrializing animal agriculture. American Journal of Agricultural Economics. Volume 77:5 (December).

Reimund, D., J. Martin, and C. Moore. 1981. Structural change in agriculture: The experience for broilers, fed cattle, and processing vegetables. Washington, DC: USDA, Economics and Statistics Service, Technical Bulletin Number 1648, April.

Rhodes, V. 1995. The industrialization of hog production. Review of Agricultural Economics. Volume 17: 107-118.

Ritchie, B. 1995. High level of nitrate in springs. Gainesville (Florida) Sun. 12 April.

Roenfedt, S. 1995. Save money now. Dairy Herd Management. 32,8: 26-28.

Sauber, C. 1992. Programmed to succeed: Florida producers optimize net profits to stay in the top 25 percent. Dairy Herd Management. April.

Schmid, A. 1987. Property, Power and Public Choice: An Inquiry into Law and Economics. New York: Praeger.

Schnitkey, G., M. Batte, E. Jones, and J. Botomogno. 1992. Information preferences of Ohio commercial farmers: Implications for Extension. American Journal of Agricultural Economics. Volume 74,3: 486-496.

Smith, K. 1995. Environmental issues from an economic perspective. In Increasing Understanding of Public Problems and Policies - 1994, S.A. Halbrook and T.E. Grace (eds.). Oak Brook, IL: Farm Foundation.

Smith, K., and P. Kuch. 1995. What do we know about opportunities for intergovernmental institutional innovation: Policy issues for an industrializing animal agriculture. American Journal of Agricultural Economics. Volume 77,5: (December).

Smothers, R. 1995. Waste spill brings legislative action: North Carolina still loves hog farms. New York Times. 30 June.

Southwestern Public Service Company. 1993. Cattle-feeding capital of the world: 1993 fed cattle survey. Unpublished report. Amarillo, Texas.

Sporleder, T. 1994. Assessing strategic alliances by agribusiness. Canadian Journal of Agricultural Economics. 42: 533-540.

Stalcup, L. 1994. A friendlier attitude: Texas dairies see less hostility from state agencies, environmentalists. Dairy Today. September.

Stalcup, L. 1992. Big stick in Texas. Dairy Today. August.

Strange, M. 1987. Family Farming: A New Economic Vision. Lincoln, NB: University of Nebraska Press.

Sweeten, J. 1993. Livestock and poultry waste management: A national overview. In National Livestock, Poultry, and Aquaculture Waste Management: Proceedings of a National Workshop. St. Joseph, MI: American Society of Agricultural Engineers.

Texas Agricultural Statistics Service. 1994. Texas agricultural statistics, 1994.

Van Arsdall, R., and K. Nelson. 1985. Economies of size in hog production. Technical Bulletin Number 1712. Washington, D.C.: USDA/ERS.

Van Horn, H.H. (editor). 1995. Nuisance concerns in animal manure management: Odor and flies. Proceedings of a Conference held March 21-22, 1995. Florida Cooperative Extension Service, University of Florida, Gainesville, FL 32611.

Warrick, J., and P. Stith. 1995 The power of pork. Raleigh (North Carolina) News-Observer. 19 February.

Westgren, R. 1994. Case studies of market coordination in the poultry industries. Canadian Journal of Agricultural Economics. 42: 565-575.

Wright, J. 1995. Agriculture and water quality in central New York's Finger Lake region: Regulatory versus voluntary programs. In Animal Waste and the Land-Water Interface, K. Steele (ed.). New York: Lewis Publishers.

Zimet, D., and T. Spreen. 1986. A target motad analysis of a crop and livestock farm in Jefferson County, Florida. Southern Journal of Agricultural Economics. pp. 175-185.

9

Privatization of Crop Production Information Service Markets

Steven A. Wolf

Within the agricultural sector it is increasingly clear that who provides what information to whom under what conditions will significantly determine the ecological, economic, and social performance of production systems. One of the key elements in the changing pattern of information flow involves private sector crop consultants assuming greater authority and responsibility in producers' farm and field level management, particularly with respect to fertilizer and pesticide decision-making. Farm chemicals represent an important component of variable production costs for farmers and impose significant economic, ecological and environmental risks on society (NRC 1989; 1993). Dependence of agriculture on farm chemicals and the controversy surrounding this dependence imply that distribution of managerial authority and the sophistication of decisions developed through such management arrangements are important topics of study.

Crop consultants provide decision-support services to farmers. The two classes of consultants addressed here are dealer-based consultants, farm advisors working for farm chemical dealers—firms that sell fertilizer and/or pesticides to farmers—and independent crop consultants—individuals that provide informational services for a fee and are not affiliated with product sales. These two information sources represent the leading private sector information service providers influencing farm chemical use.[1]

The author gratefully acknowledges Pete Nowak's contributions to this research and Rachael Goodhue's comments on an earlier version of this chapter. This research was conducted, in part, through a grant from the USEPA Agriculture Policy Branch (agreement CR822762-01-0).

ISBN 1-57444-104-3/98/$0.00/$.50

This paper focuses on the development and structure of consulting markets organized around information-based farm chemical management services—analytic and diagnostic practices that use spatially and temporally specific data to inform fertilizer and pesticide use decisions (Table 1). Because of their environmental and economic benefits, these waste minimizing fertility and pest management techniques have been promoted by natural resource conservation interests, Extension agents, and independent crop consultants for many years (i.e., integrated pest and nutrient management). This paper addresses the rationale for and mechanism through which dealers are entering into agronomic service markets and the relationships between dealers and independent crop consultants. I de-

Table 1. Examples of Information-Based Farm Chemical Management Services

Service	Description
Soil sampling	Taking soil samples from farm fields, generally for purposes of fertility analysis. Intensity of sampling varies spatially and temporally. Sampling procedure highly variable (e.g., depth and handling of samples). Sophistication varies from a single composite sample for a set of fields to grid sampling using Global Positioning System technology (GPS).
Fertility analysis/fertility recommendations	Testing of soil samples for plant available nutrients and pH. Laboratory procedures and assay techniques variable. Alternative "philosophies" (yielding disparate recommendations) guiding interpretation of lab results include mass-balance, cation saturation ratio, and sufficiency.
Nitrate-nitrogen testing (pre-plant/side dress)	Soil test for residual nitrogen available to the plant. Test results inform pre-plant (PPNT), topdress or presidedress (PSNT) nitrogen application (Meisenger et al. 1993). Depth of sampling and frequency varied.
Plant tissue testing	Leaves/petioles analyzed during early and mid-season to determine nutrient needs/deficiencies. Can be combined with soil testing to assess whether soil nutrients are reaching the crop. Used both prophylactically and as a diagnostic tool. Frequency varied.
Manure analysis	Testing of animal manure for nutrient content (N-P-K). Variable with respect to frequency. Method of manure storage affects rate of denitrification making nutrient crediting inexact.
Manure spreader calibration	Estimates quantity of manure of particular quality applied per acre by a specific spreader traveling at a particular groundspeed. This datum is used with manure analysis data to calculate nutrients added per acre through manure application.
Nutrient management budgeting/planning	Development, maintenance and implementation support of a nutrient budget, integrating information on nutrients added to each land management unit from chemical fertilizer, manure, legumes, sludge or other sources. Plans vary in their detail, accuracy, comprehensiveness, and ease of understanding.
Pest scouting	Monitoring pests (insects, disease, weeds, etc.) to determine if infestation has reached an economic threshold. Training of scouts, sampling procedures, application of thresholds, and frequency and duration of field visits are key sources of variation
Pest/Pesticide record keeping	Log of pesticide use. Required in some states for restricted use pesticides. Information content variable, and logs use as a management tool variable. Records can be used for production planning, pesticide mode of action rotation (pesticide resistance control), litigation, economic analysis, informal experimentation.
Pest prediction models	Use of quantitative models or expert/decision-support systems to make predictive estimates based on empirically derived field data. Models are highly varied with respect to level of detail of both front-end data requirements (input) and as well as specificity of recommendations (output).

scribe contemporary developments in key institutions involved in consulting markets and report findings based on a series of community-level case studies. These empirical observations inform our understanding of the emerging crop consulting industry and serve to frame questions as to the economic, environmental and social implications of greater reliance on private sector information providers in farming systems.

EMERGENCE OF INFORMATION-BASED COMPETITION IN AGRICULTURE

The implications of development, control and application of information, has emerged as a contemporary focus of inquiry among academics, economic actors, and policy makers due to a convergence of factors operating at three levels: the macrostructural-level, agro-food system-level, and farming system-level.

Macrostructural factors condition the development of agriculture. Overarching, cross-cutting issues of note include socio-political intervention in production systems focused on environmental accountability and consumer protection (Buttel 1992) and public fiscal austerity. Each of these factors indicate a change in the nature of the relationship between agriculture and society. Public sector interest in agricultural information is likely to be increasingly focused on health and safety regulation, natural resource management, and basic rather than near-market research. Additionally, agricultural research is increasingly privatized—i.e., funded and conducted by private firms (ERS 1995).[2] In these respects, agriculture is becoming more like other industries as the "social contract" between agriculture and society is eroding (Hebert 1995). As governmental regulation of agricultural production and trade declines, as signaled by movement away from commodity programs in the 1996 Farm Bill and multinational trading agreements such as NAFTA and GATT, market forces are expected to become more important. Development and application of superior information is expected to be a primary strategy of firms engaged in highly competitive markets.

Developments within the agro-food sector pose risks and opportunities for farms and non-farm agribusiness. Specifically, structural change—concentration, vertical coordination, and vertical integration—has led to production units which are larger, specialized, and more interdependent. Under these conditions, logistical control and resource allocation decisions warrant a sophisticated management approach. Additionally, the inelasticity of demand for basic agricultural commodities is spurring interest in development, preservation and enhancement of specific product attributes and development of new products (see Zilberman et al.

this volume). Traditional markets are becoming fragmented, for example, organic grains, colored cotton, and specialty fruits and vegetables. Differentiation processes and participation in niche markets are necessarily information-intensive as production, marketing, and quality assurance require more exacting attention. Technological change (e.g., biotechnology, telecommunications, internet, computerized data bases, bar-coding, precision farming) has supported differentiation of agricultural products and increased managerial oversight capacity. As agriculture moves toward "just-in-time" and precision-oriented systems of industrial manufacture, information is emerging as the key to success.

The third level at which change must be considered is that of local farming systems, the focus of this paper. The structure and performance of information markets at the community-, farm-, and field-scales are particularly important as this is where the environmental interface is most direct (i.e., soil, water, air, wildlife, humans). Mike Boehlje's assertion (this volume) that firms are now demanding "context-specific and decision-focused" information, and therefore we are witnessing growth in private sector consulting activity, is a key insight. In theoretical or general terms, analysists have begun the process of identifying the direction and significance of information market trends. However, at the community- and farming system-level we have extremely limited knowledge as to local processes through which information is developed, distributed, and applied. Who will supply farmers with customized information? What factors determine the quality of fertilizer and pesticide management recommendations? Will all farmers have access to information of similar quality? Under what conditions is conflict of interest—bias arising from firms recommending products that they sell—a problem? What will be the role of Cooperative Extension as consultants become more important? These questions highlight the significance of consultants becoming more fully integrated in farmers' decision-making processes.

The remainder of this paper is divided into four sections. In the first section of the paper I identify environmental and economic implications of information-based farm chemical management services and go on to discuss contradictions associated with enhancement of off-farm involvement in on-farm fertilizer and pesticide management. In sections two and three I describe conditions driving and constraining development of crop consulting markets. The second section focuses on developments and strategies pursued by national organizations representing dealers and independent consultants. In the third section empirical results of a series of community-based case studies are reported. Findings are based on 122 personal interviews with farmers, consultants, and Extension personnel in three states. In the final section of the paper, I discuss factors affecting the quality of farm chemical recommendations available to farmers with particular emphasis on consultants' potential conflict of interest.

POLITICAL ECOLOGY OF FARM CHEMICAL WASTE MINIMIZATION

The enhancement of farm- and field-level decision making processes through application of temporally- and geographically-specific information may mitigate negative environmental impacts associated with fertilizer and pesticide use and misuse (Fuglie and Bosch 1995; Robert et al. 1993, 1995; Swinton and King 1994). In theory, through application of higher quality information, the correlation between variance in biophysical setting and variance in management of crop inputs will be enhanced (Wolf and Nowak 1995).

In discussions of information-intensive agriculture, also referred to in some contexts as site-specific or precision-farming, information is often characterized as a "substitute" for farm chemicals. This is inaccurate in the sense that farm chemical use will not necessarily go down as a result of application of analytic and diagnostic procedures. In fact, in some cases input use will rise if strict economic rules are applied to input decision making.[3] Regardless, the volume of inputs is only one factor of importance. Due to heterogeneity in the spatial and temporal distribution of natural pollution attenuation capacity and human and natural resources, risks associated with farm chemical inputs—the likelihood of a negative impact and the significance of that impact—are heavily influenced by when, where, how and what types of inputs are applied. Information-intensive approaches imply enhanced environmental performance not through absolute source reduction but through more accurate targeting of input investments. By matching crop needs to crop inputs at increasingly fine spatial and temporal scale, a higher percentage of farm chemicals applied to crop fields will serve their agronomic purpose and a lower percentage will miss their intended target, resulting in less pollution.

In addition to the environmental and ecological benefits associated with information-based agrichemical management, as waste minimizing technologies, input use efficiency and economic productivity are expected to rise. In theory, both under-investment (i.e., cases where marginal investments in inputs would produce acceptable rates of return) and over-investment (i.e., cases where marginal investments in inputs feature negative returns) can be avoided. More precise allocation of inputs provides a mechanism to reduce production costs, enhance yields, and product quality, and manage risks. The reinforcing nature of economic and environmental opportunities represented by these farming practices make development of agrichemical management service markets a prime example of a market-based approach to agroenvironmental management. Market-based strategies are consistent with a shift in patterns of government intervention and are closely related to deregulation and privatization trends (Bonanno 1990).

Growth in dealers' and independent consultants' roles in on-farm decisions is occurring at a time when public sector institutions, most notably USDA Extension and the Natural Resources Conservation Service (NRCS), are assuming diminished roles with respect to development and direct delivery of agricultural information (Bennett 1994; Buttel 1991; Keystone Center 1995). The combination of these two trends—more private and less public involvement in farming systems—represent privatization of crop production information services. A trend toward privatization of extension (Rivera and Cary in press; Rivera and Gustafson 1991), now observed in the U.S., is more advanced in other countries, notably the U.K. (Bunney this volume) and New Zealand (Cary this volume; Hercus 1991). While in the case of Britain and New Zealand an official program of privatization of extension has been enacted, here in the U.S. no formal transfer is underway. What is observed in the U.S. is relative growth in private sector authority as compared to that of public sector institutions.

Public sector institutions continue to exert influence over farm-level activity and field-level decision making. This influence, however, is increasingly indirect. In addition to macro policy tools such as fiscal, trade, and land use policies, access to producers' decision making is maintained through agricultural research and technology transfer. Additionally, public sector authority is exercised through regulation (e.g., FIFRA and state regulations), quasi-voluntary initiatives (e.g., Cross Compliance), and voluntary financial and technical assistance programs (e.g., NRCS and Extension cost-sharing and educational programs). While public institutions still play a pervasive role, their influence over field-level production activities is diminished relative to historical levels.[4] Diminished roles for public sector agricultural institutions can be attributed to public sector fiscal austerity, depopulation of agriculture, and declining political and economic power of farm interests, generally (Buttel 1991; Smith 1996). However, such an explanation fails to incorporate the interests of agribusiness.

While there is a solid economic and environmental logic to application of waste minimization principles to farm chemical management, and the timing of diminished public sector agricultural Extension can be rationalized through consideration of the historical shifts in the significance of agriculture relative to the non-agricultural economy, there are reasons to apply a more critical and explicitly sociological analytic perspective. These technologies and the production system of which they are a part are recognized as expressions of the social organization that generated them and that is produced and reproduced by them (Busch et al. 1991). The crop consulting industry is premised on the commodification of information—the process of establishing property rights over goods for the purpose of controlling rents (Goe 1986; Wolf and Wood forthcoming). Enhanced roles for off-farm firms in on-farm decision making is consistent with a long run pattern of "appropriation and substitution" (Goodman et al. 1987) that underlies the industrialization of agriculture and concentration of political and economic power. As components of production previously derived from on-farm resources are replaced with corresponding inputs sourced from off-farm industrial outlets, the significance of atomistic farmers' contributions to the production process is marginalized relative to that of centralized agribusiness. While the impacts of industrialization are not clearly understood, environmental and ecological hazards associated with large scale production systems, barriers to entry for new firms, concentration of political and economic authority, and rural economic problems in agriculturally dependent regions merit scrutiny.

Further, as we have argued elsewhere in the context of precision farming technology (Wolf and Buttel this volume), much of the motivation underlying agrichemical waste minimization innovations can be linked to mainstream agriculture's growing need to deflect mounting environmental criticism, rather than producer-derived demand. Development and supply of farm- and field-level efficiency enhancing services would appear to have as much to do with agribusinesses' need to reinforce the legitimacy of high-input agricultural production systems as it does with farmers' demand for additional support. Technologies that symbolize a commitment to and may result in an incremental reduction in externalities associated with chemical fertilizer and synthetic pesticides conserve and strengthen the status of these industrial inputs. These developments can be seen as part of a process of restructuring of agricultural practices so as to redress a set of widely recognized problems while simultaneously protecting and advancing the industrial structures, investments, and institutional arrangements premised on these practices.

COMPETING OUTLETS: AGRICHEMICAL DEALERS AND INDEPENDENT CROP CONSULTANTS

Here I describe developments within the agricultural input supply sector and national institutions representing consultants. These observations contribute to understanding of the structure and developmental trajectory of crop consulting markets. Institutional activities reveal strategic interests. The outcome of these strategic initiatives are likely to have important local impacts within farming systems.

Agrichemical retail dealers are moving quickly to enhance their presence in information markets and formalize their role as service providers within cropping systems. While communication of product related information between farm chemical supply dealers and farmers has traditionally been informal (not treated as a business function separate from product sales) and therefore widely under-recognized by analysts, these firms play a critical role with respect to the information flows in farming systems (e.g., Ford and Babb 1989; Schnitkey et al. 1992). As components of farm chemical distribution systems, custom applicators, and sources of information these firms play significant roles in determining the extent and intensity of agrichemical use and geographic and temporal patterns of adoption of efficiency enhancing innovations.

A significant factor which has motivated growth and formalization of dealers' information-based service offerings is the need to identify new profit centers and sustain existing business roles at a time when many local agribusinesses are failing (Ginder 1992). The maturation of the industry—many suppliers selling similar products, at similar prices with little room for introduction of new products—has reportedly greatly reduced fertilizer and pesticide profit margins (Peitscher 1991). Estimates suggest that in 1980 there were 18,000 fertilizer and pesticide dealers. It is suggested that 20 to 30 percent of the estimated 12,000 U.S. agrichemical supply businesses in operation in the early 1990s will go out of business over the next 10 years (OTA 1990). Within this competitive environment dealers are expected to move to secure their relationships with clients and develop spin-off markets in farm chemical services. Service sales are potentially profitable enterprises, as well as a means by which firms can retain and attract new customers for their traditional products. As agrichemical supply firms struggle to maintain and expand their market share at a time when low product margins have eroded price differentials between firms, services are expected to become the differentiating feature between competing dealers (Gannon and Ginder 1992; Hoffman 1993; Keeney 1991; Lambur et al. 1989; Wolf and Nowak 1995).

Strengthening and broadening of dealer-farmer relations has been strongly advocated by professional associations representing agrichemical retail dealers. Dealer associations are promoting an active role for agrichemical dealers with respect to information provision and service development in order to project an environmentally sensitive image for their industry, diversify earning opportunities, and protect their members' position within rapidly restructuring agricultural production and marketing systems (Peitscher 1991; Sine 1990).

Farmers' sources of agrichemicals and agrichemical management information are changing. Vertical integration within agrichemical manufacturing, distribution, and retailing sectors is proceeding. Agrichemical manufacturers and wholesalers are expanding their presence in retail markets by selling discounted, bulk products directly to larger farms. Also, independent crop consultants are becoming increasingly popular fixtures in a wider variety of cropping systems. Each of these trends may serve to marginalize local retailers. By establishing themselves as providers of information and production services, dealers are attempting to add value to their products and retain and expand their market share. As observed across many segments of the U.S. economy, modern production, marketing, and distribution systems frequently "cut out the middle-man." Service development can be viewed, in part, as dealers' strategy to avoid such a fate.

Evolution of a highly visible and voluntary professional certification program—the Certified Crop Advisor (CCA) program—is testament to the agrichemical industry's interest in participating in crop consulting markets. This program also testifies to the competitive dimension of relations between national organizations representing independent consultants and dealers. Certification and other forms of accreditation serve as symbols of quality in the marketplace and as a mechanism through which industries can develop and project a professional image. These programs also represent a means through which professions can police themselves and create barriers to entry for other firms. While individual states have long administered licensing programs for pesticide use consultants, there has been substantial recent growth in private sector accreditation programs.

In 1991 in an effort to raise the performance standards of the industry and to address the specter of potentially conflicting interests faced by individuals that consult on the use of products that they sell, the pesticide and fertilizer industry, USDA, EPA, and the American Society of Agronomy initiated the CCA program. The CCA program was a direct reaction to USDA's decision to accord relative autonomy to independent consultants while requiring USDA oversight of water quality management plans written by dealers who were perceived as potentially biased (Dysart 1994).[5] The program consists of a national examination, a state examination, a continuing education requirement, and adherence to a code of ethics, as well as fulfillment of a range of eligibility criteria. The popularity of the CCA program is portrayed in Figure 1. As the chairman of the Board of Directors has stated, the CCA program is not explicitly a "dealer program", as anyone is eligible to apply for certification (Watts 1995). However, the vast majority of participants are employees of agrichemical dealerships. The large number of individuals taking the CCA examination, introduction of CCA in Canada, and general support from industry leaders for development of a service orientation within dealerships is strong evidence that the agrichemical industry is positioning itself to justify its actions in the face of mounting environmental criticism and to capitalize on what is expected to be a large and growing crop consulting market.

Also in 1991, the National Alliance of Independent Crop Consultants (NAICC), the largest professional society representing independent consultants and private researchers, founded the Certified Professional Crop Consultant (CPCC) program. CPCC requires examination, continuing education, and adherence to a code of

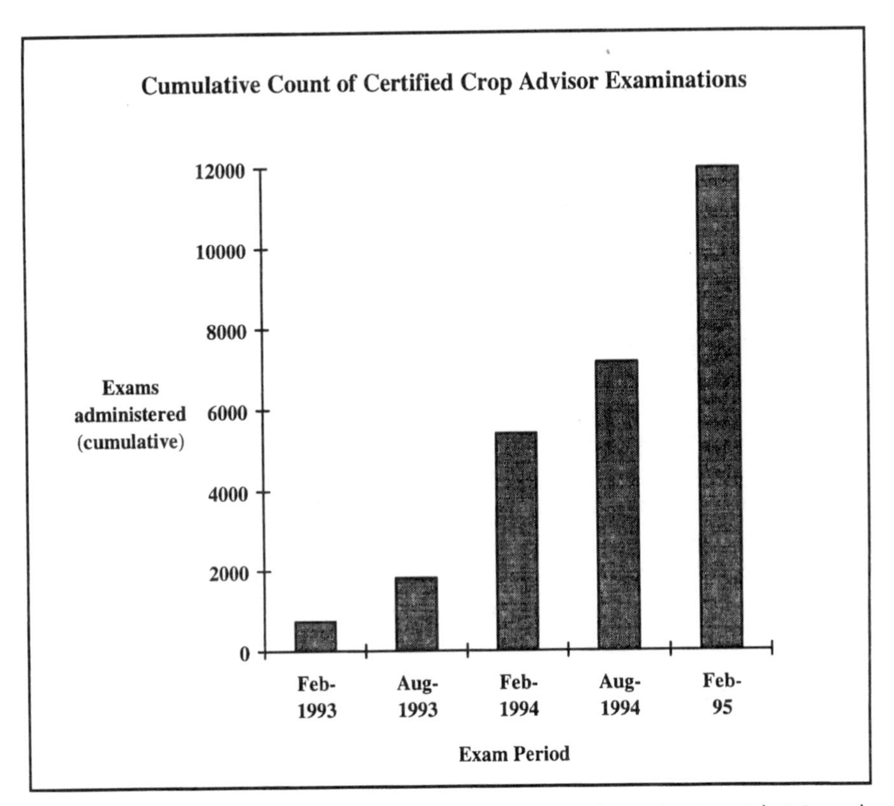

Figure 1. Cumulative Total of Certified Crop Advisor National Examinations Administered (Source: American Society of Agronomy)

ethics. CPCC-Independent (CPCC-I) is an additional designation reserved for consultants who have no connection to product sales of any kind. Independent consultants benefit from the perception that they are free of any conflict of interest stemming from disincentives to reduce farmers' input purchases. Key elements of the two certification programs are compared in Table 2. Differences between the programs on most criteria indicate the independent consulting industry's efforts to distinguish itself through more rigorous standards.

While independent consultants have been providing professional services to farmers for over forty years (Lambert 1995), their participation has been limited to specific commodities, for example cotton, rice, fruit, and vegetable production systems. In food and feed grain production systems, prominence is more recent

Table 2. Comparison of the Certified Professional Crop Consultant (Independent) CPCC(I) and the Certified Crop Advisor (CCA) Program.

Criterion	CPCC/CPCC-I	CCA
Education	BS or BA in agriculture, pest management or biology. Non-agr., 4 year degree eligible if 10 years experience	No minimum
Experience	Ph.D. requires minimum 4 years consulting experience, MS requires 5 years, BS requires 6 years	High school requires minimum 4 years consulting experience, Associates Degree 3 years, 2 years with BS or higher degree
Examination	State CCA examination or state licensing exam, plus a case study narrative.	National and state CCA exam
Continuing education units (CEU)	36 hours per year of approved educational programming	20 hours per year of approved educational programming
Code of Ethics	NAICC code of ethics	ARPACS code of ethics
Independence from product sales	No/Yes	No
Professional references	3	2

(Bocher 1990). A 1993 survey sponsored by the NAICC estimated that members of the profession now provide consulting services on 53 percent of cotton acres, 53 percent of vegetable acres, 21 percent of corn acres and 13 percent of soybean acres. Overall, the study estimated that independent crop consultants provide analytic and diagnostic services on 16 percent of U.S. cropland (Nowlin 1993). As represented by growth in NAICC membership (Figure 2), independent crop consulting is apparently a steadily growing profession.

In an era of structural and technological change and one in which environmental accountability is increasingly important, information requirements in farming systems are evolving rapidly. Competition in information markets is a driver of activity at the level of national organizations as well as among individuals engaged in servicing farmers. Dealers are observed to be motivated by economic and political factors to participate in service markets. Independent consultants are observed to be taking steps to protect and expand their position in agricultural systems.

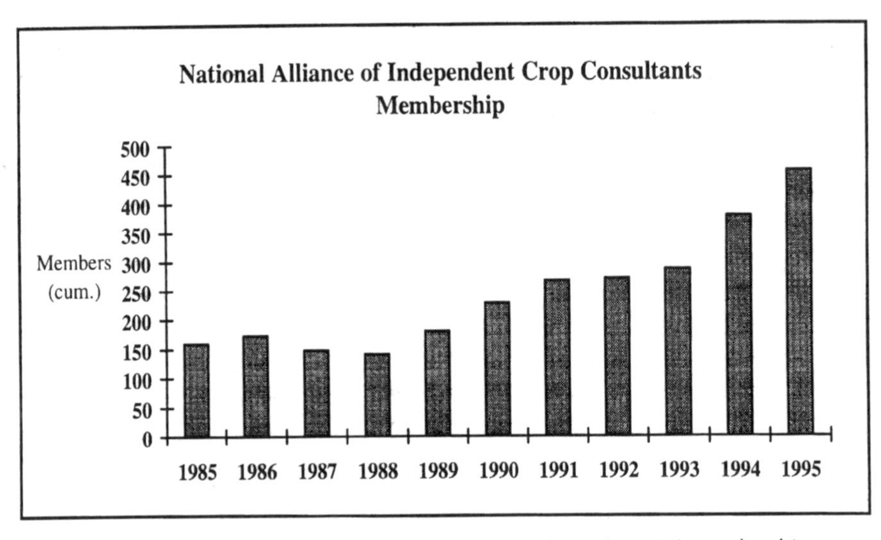

Figure 2. National Alliance of Independent Crop Consultants (NAICC) Membership (Source: NAICC)

CONSULTING SERVICES AND FARMING SYSTEMS

In this section of the paper I review findings of a series of six community-level case studies. The case studies focused on production and consumption of information-based farm chemical management services. Interviews were conducted in two production regions in each of three states; corn production sites in a Midwestern state, cotton production sites in a Mississippi Delta state, and mixed crop production sites in a mid-Atlantic state. Developments within and interaction between four subsectors are analyzed: i) the farm sector, ii) public sector Extension, iii) the farm chemical supply industry, and iv) the independent crop consulting industry. These four components of agricultural systems are regarded as key determinants of the structure and performance of crop production information markets.

The two sites in each state were selected for inclusion in the study according to level of competition within local agrichemical supply and information markets. Within each state a "higher competition site" was selected, as defined by an area with a relatively large number of dealers and independent consultants, and a "lower competition site" was selected representing an area with a relatively small number of product and information outlets. It should be noted that areas of higher competition generally feature more intensive agricultural production. Competition between local agribusinesses reflects the relative size of these input markets, intensity of land use and farm chemical input use, and the nature of agriculture in these areas (Wolf 1995). Competition between local agribusinesses is an integral characteristic of the production system, not an isolated attribute. Regardless of asymmetry between high and low competition sites, this study design facilitates analysis of the impact of competition within a community on developments in local service markets. In an era in which market forces are of increasing importance relative to state intervention and direct regulatory control, the role of competition in generating social benefits is of heightened relevance.

As discussed above, consolidation and restructuring in the agrichemical supply sector is proceeding rapidly. Aspects of this study are premised on the hypothesis that in higher competition settings, dealers and consultants are actively investing in service development as a means to differentiate themselves from competitors and secure their positions in farming systems. A related hypothesis—and one not directly testable through application of these data—is that service development under conditions of high competition results in high quality crop management recommendations that increasingly improve crop input efficiency thereby enhancing the economic and environmental performance of cropping systems. While understanding of some aspects of information quality are informed by these data, our ability to make statements regarding the site-specific economic and environmental implications of certain analytical and diagnostic procedures remains largely undeveloped (see Wolf 1996).

Data derived from 122 personal interviews with farmers, Extension, dealers and independent consultants are combined with a set of broader observations on emerging trends in agriculture. The heterogeneity of agricultural production systems within and across communities makes synthesis challenging and often misleading. As these findings are based on limited data, this summary should not be regarded as a comprehensive portrait. These observations are communicated in order to highlight a set of important research and policy issues.

Farm sector

Farmers are seeking support from consultants due to a range of internal and external factors including structural, economic, technical, institutional, environmental and agroecological reasons. Relationships with consultants are focused on decision-support aimed toward raising yields, enhancing quality, lowering costs, and/or managing risks. The trends of increasing acreage per farm and greater farm specialization serve to enhance farm sector demand for off-farm production assistance. As crop production has become increasingly sophisticated and capital intense, use of technical specialists who posses the requisite skills and tools makes increasing economic sense. The high cost and low quality of hired labor (and its reported unavailability in some regions), patterns of allocation of on-farm human capital, and the opportunity costs of learning and executing analytic and diagnostic techniques lend appeal to subcontracting for all or part of the fertilizer and/or pesticide management components of crop production. Liability considerations focused around worker protection standards, natural resource degradation, and damages to neighbors' property support outsourcing farm chemical tasks. In some senses farmers are moving into the role of "general contractor" in which they choreograph the activity of various subcontractors.

While efficiency objectives are a powerful driver of closer relations between farmers and consultants, farmers also seek to organize their farming operations to fit their personal interests and ambitions. Health and safety concerns are contributing to some farmers' decision to rely on off-farm firms to service their agrichemical requirements. Consultants are providing soil sampling and pest scouting services on some farms because farmers can afford to pay others to do work perceived as boring or exhausting. Additionally, some farmers satisfy their intellectual curiosity through working closely with consultants on technical or agroecological aspects of farming.

Technological change

The pace with which technologies are emerging and new information is becoming available serves to make knowledge obsolete in short order. Staying up to date with the latest production-oriented research in highly specialized areas, for example weed control in corn or growth regulators in cotton, represents a time-consuming activity. The opportunity costs of developing and maintaining on-farm capabilities to manage and execute all facets of crop production are rising to the point where outsourcing services is increasingly attractive. Further, by virtue of their labor requirements, cost, scale, requisite infrastructure and/or esoteric skill requirements, emerging technologies, for example the cluster of technologies associated with precision farming and biotechnology, increasingly demand off-farm support. Incongruence between on-farm human capital and the requirements of emerging technologies is exacerbated by the high average age of farm operators.

The linked nature of component technologies (e.g., seed-herbicide combinations) and the marketing of agronomic packages lead to close association between farmers and consultants. Agrichemical dealers have traditionally offered informal insurance to farmers who purchase an integrated set of products (seed, fertilizer, pesticides) and faithfully follow the dealers' management guidelines. Farmers who purchase and apply sets of products at the recommended rates at the recommended times in the recommended fashion are eligible for partial reimbursement or in-store credit in the event of crop loss. By accepting a certain level of responsibility for the crop's success through extending product performance guarantees, dealers are able to exercise a level of control over how their products are used.

Environment

The environment is both a legally protected resource creating requirements on farmers and a dynamic component of cropping systems requiring adaptation on the part of farmers. Consultants are involved in shaping responses to both of these challenges. Governmental regulation and environmental liability add complexity, cost and risk to production. Concern for environmental quality, more stringent standards of accountability, and greater penalties for non-compliance serve to motivate farmers to seek additional information to guide management. While environmental protection is frequently cited as adding complexity to production, environmental change also leads to increased demand for information. Past management behaviors are increasingly recognized as triggering ecological pressures that require changes in production systems (NRC 1992). For example, soil erosion, compaction, and reduction in organic matter content, infiltration, and water holding capacity have contributed to the popularity of conservation tillage and chemical weed control. Pesticide resistance and secondary pest outbreaks are motivating changes in pesticide use, cultural practices and land use patterns. Initiating such changes requires technical support. The status of consultants is boosted by farmers' needs to respond to changes in the ecological setting in which production occurs.

Farm sector diversity

The previous discussion reviewed many features of contemporary agriculture contributing to integration of consultants into on-farm production decisions. In some form, each justification for expansion in demand for information was observed in the case study data. However, not all farmers were moving to source more information or to source better information. In some cases farmers were not reacting to any of these factors and exhibit no intention of increasing their information gathering. Some who had employed a given technique (e.g., soil nitrate or tissue testing) or relied on a consultant's services in the past have decided to discontinue specific practices based on their experience. As would be expected, this cross sectional analysis indicates pronounced diversity in attitudes and relationships with consultants across individual farms.

Generally speaking, relative to lower competition sites, farmers in higher competition regions consumed more analytical and diagnostic services. In higher competition regions, farmers were more likely to employ an independent consultant and to pay more for information. In these markets farmers were more likely to "triangulate" on recommendations through consideration of a second opinion. As would be expected, these farmers were more knowledgeable, in a relative sense, and had more experience concerning integrated crop management. Farming practices in higher competition regions featured more frequent and detailed soil sampling (Global Positioning Satellite based, grid soil sampling in the case of the Corn Belt site), more intensive pest scouting and more sophisticated record keeping. In higher competition sites, interaction between farmers and consultants was observed to be more extensive and intensive.

While the general trend is toward more extensive and intensive use of consultants and information-based techniques to guide input management it is important to note that contemporary procedures for determining if, when, how and how much of which inputs to use where are not rigorous. National rates of adoption of soil testing and pest scouting, and, perhaps more importantly, the intensity or sophistication of field-level application of these management practices, indicate that farm chemical management is significantly not site-specific nor information-intense (Vandeman et al. 1994).

Farmers' sensitivity to information quality as measured by patterns of information gathering and willingness to pay for consulting services were observed to be highly variable; variable by region, by farmer, and by specific decision type. Farmers were aware of the potential for a conflict of interest to effect advisors' recommendations. Many expressed that they had personal experience with biased information. Farmers' strategies for dealing with potentially biased information were highly variable. Some farmers triangulated to arrive at a decision by gathering as many as three opinions (e.g., dealer, independent consultant, and extension service) on a single pesticide treatment decision, while others sprayed when the dealer recommended spraying. Farmers' ability to pay for high quality information and their level of concern regarding bias are significant determinants of the pace and nature of development of consulting markets.

The status of consulting markets is an important determinant of the sophistication of farmers' management, specifically consumption of formalized analytic and diagnostic procedures. However, other information channels—family, friends, media, etc.—and economic, environmental, and social variables are obviously important elements of the overall context in which input use decisions are made. Use of commercial service offerings and the quality of those services within a community are only one of a larger set of factors likely to determine input use efficiencies.

Public sector Extension

The three regions, Corn Belt, Mississippi Delta, and Mid-Atlantic, exhibited the process of privatization at three points in time. In terms of the role of locally-based Extension agents, privatization is largely complete in the Corn Belt state (no county agricultural Extension agents and limited regional programming), actively progressing in the Mississippi Delta state (county Extension services discontinuing), and beginning to emerge in the Mid-Atlantic state (private service markets just forming).

A privatization gradient can be observed across higher and lower competition counties within each state. In higher competition sites Extension was less active and relied upon, independent crop consultants were more active, and dealers were more fully engaged in service delivery and promotion. The level of development of private service markets was observed to be negatively correlated with farmers' use of Extension programming. This finding is consistent with Extension's interest in not competing with the private sector and in allocating resources to those areas where private sector investment is inadequate (Bennett 1994).

Public sector institutions remain important components of agricultural production information markets as they are the original suppliers of much of the research that undergirds consultants' recommendations. In more direct terms Extension and other public agencies shape interaction between private consultants and farmers through publications, education, demonstration, and certification. Extension also serves as an "incubator" for future consultants through provision of a variety of training opportunities.

Although somewhat variable from site to site, Extension generally did not enjoy close relations with dealers in the study sites. With notable exceptions, many dealers expressed low regard for the agronomic knowledge of local Extension personnel. Based on my observations, independent crop consultants utilize and respect Extension personnel to a greater degree than do dealers. University Extension specialists were generally well respected but not heavily involved in working one-on-one with dealers and independent consultants.

Diversity in Extension

Farmers' use of and regard for Extension services were highly variable. The information from publications and meetings sponsored by Extension were valued by a minority segment of farmers in each of the six study sites. These sources of information, like the occasional recommendations offered by a university specialist on the occasion of an obscure pest outbreak, were regarded by the majority of farmers as ancillary sources of information.

In the Corn-Belt state, Extension had recently been reorganized, largely eliminating its position as a source of direct farm chemical recommendations in both study sites. In the Mississippi Delta state Extension was also recently reorganized. In the higher competition site where pest scouting in cotton is a firmly established private sector industry, Extension was not a major source of crop production advice and played a minor role in validating private sector recommendations for those growers who wanted to verify the information provided by their consultants. In the lower competition site in the Mississippi Delta state where cotton crop consulting was a less established feature of production systems, the county agent administered a county-based pest scouting service for a fee. This service is being phased out over the next few years as the number of farms declines and dealers and independent consultants expand their services.

In the Mid-Atlantic state, Extension was also recently reorganized. In this case the change was toward more county-based agents and expansion of programming. Extension provided nutrient management planning assistance and facilitated integrated pest management and pest scouting programs. The situation in the Mid-Atlantic state is regarded as an aberration relative to national trends due to the unusual circumstances and funding resources associated with the Chesapeake Bay clean up effort.

The field presence of Extension is shrinking. It is clear that the county agent is not and will not be a major source of production information and farm chemical recommendations. County agents now have increasing responsibility for youth programming, economic development, community facilitation services and other non-agricultural production related activities. In many states, the county agricultural agent position no longer exists and several counties now share a pool of technical specialists. This strategy away from organization of Extension according to geographic boundaries and towards disciplinarily organized Extension is consistent with the rising volume, specificity, and complexity of information. In keeping with this trend, university based Extension specialists are increasingly relied upon by farmers. When a farmer wants a question answered by extension, the person called is increasingly a technical specialist housed at a regional office or the university, rather than a county agent.

Agrichemical dealers as information providers

Accelerated restructuring of the agrichemical supply industry is contributing to the rapid entry of dealers into crop consulting markets. Three significant developments related to structural change are discussed: vertical integration, market concentration, and one-stop shopping. I then address dealers' roles in service markets.

Vertical integration

As mentioned earlier in this paper, farm chemical wholesalers (as well as manufacturers) are expanding their operations in the retail sector. Within five of the six study sites wholesalers' bold entrance into retail markets was observed to be a key driver of change in farming systems. These firms are expanding their retail presence by expanding their current facilities, purchasing local facilities, building new outlets, and in some cases, recruiting local people with good reputations. These firms have greater resources and are positioned to take advantage of economies of scale. They are not saddled with outdated physical plants and they have longer strategic planning horizons than locally based, independent competitors.

Concentration

Movement of wholesalers and distributors into local communities is part of a larger pattern of concentration in the industry. Many input supply firms are going out of business or merging with other companies. The trend is toward fewer, larger firms. Large dealerships are buying out smaller dealerships resulting in parent dealers operating increasing numbers of satellite facilities. In most cases newly purchased outlets remain open for business and retain many of the same employees including managerial staff, but are operated under new owners. Some satellite facilities serve as full service outlets, offering a full line of products, custom application equipment, and agronomic services. Other satellites are simply product supply depots, have no trained staff and offer no services. The number of firms in the industry is shrinking, yet the total number of outlets is changing at a different rate.

Despite the decided trend toward fewer, larger dealerships, it is likely that independent dealers will persist into the foreseeable future and will be important components of local production systems. While these businesses are not well positioned to take advantage of returns to scale on facility design, office automation, insurance, regulatory compliance investments, and marketing opportunities, they will persist through combinations of self exploitation (low returns to capital and labor), geographic isolation (e.g., poor transportation infrastructure limiting competitors' entrance into local markets), customer loyalty, and providing difficult to replace, community-specific, time-honed knowledge of local agroecosystems. Small firms are likely to fare better in minor markets less attractive to outside capital.

This is not to imply that these smaller dealers will necessarily feature outmoded technology and survive on a blend of folksiness and intuition of local agroecological factors. The key to these smaller retail outlets is retention of highly knowledgeable crop specialists who are recognized by area farmers as possessing valuable knowledge of local agricultural practice. Surviving firms will require technical expertise and access to state-of-the-art tools and information. Powerful factors driving consolidation and homogenization combined with a set of countervailing factors constraining concentration and conserving diversity are likely to contribute to development of a bimodal distribution of firms parallel to that of the farm sector.

One-stop-shopping

The last dimension of structural adjustment in the farm chemical industry mentioned here is expansion of product and service lines. Provision of one-stop-shopping is a leading objective of firms. Fertilizer companies are adding pesticides, companies traditionally dealing in dry fertilizer only are adding liquid blending facilities, and firms that sold only pesticides in the past are adding or are actively considering adding fertilizer sales. Across all of these sub-sectors, firms are developing services to accompany their product lines. Firms are diversifying and offering expanded choices and convenience to their customers. While individual firms are diversifying their functions, the overall effect is one of homogenization. As increasing numbers of firms move to carry the full spectrum of products, specialty dealers (e.g., fertilizer only) become a rarity.

Service growth

In keeping with the trend toward one-stop-shopping, dealers are attempting to offer farmers convenience and value through integration of products and services. Dealers in each of the six study sites were moving to expand analytic and diagnostic service offerings. Structural change and competition are motivating dealers to move into service sales as a means to build and conserve market share. As a manager at one dealership stated, "this industry will not be a commodity business in the future, it will be a service business." The pace of this transition is directly related to levels of competition in local markets. While the developments described here are occurring throughout the six study sites, within higher competition regions evolution of services is accelerated sharply.

Service growth within dealerships is largely focused on soil fertility analysis and pest scouting. A range of additional practices related to these core services are also offered by some dealers including plant tissue analysis, fall soil nitrate testing, pre-side dress nitrogen testing, manure analysis, and weed scouting. Much of the growth in dealer services is intensification of services traditionally provided on an informal basis, not necessarily an addition of services previously not offered. Many dealers have been willing, upon request, to take soil samples and develop fertility recommendations for a long time. Similarly, dealers engaged in much informal pest scouting by doing windshield surveys, walking fields with farmers, and riding combines at harvest. What is new is the prominence of services in dealers' approach to relating to farmers, formalization of service as an enterprise within the firm, and the intensification of services as demonstrated through investment in infrastructure (e.g., GPS, computer software, laboratory facilities, variable rate applicators, high clearance spray rigs) and personnel.

Dealers are pursuing varied strategies to enhance the knowledge, skills and professionalism of their personnel. Staff are encouraged and in many cases required to obtain CCA certification. New, more highly qualified staff are being hired. The leading agronomists in some communities are subjects of bidding wars and are hired away from one dealership by another, bringing with them their loyal farm patrons. Dealers are engaging independent consultants to provide services to farmers on a subcontract basis in order to provide more and better information to farmers. By outsourcing service capacity dealers are able to partially shield themselves from charges of conflict of interest. Alliances between dealers and independent consultants have significant implications for the structure of local information markets. Relationships between dealers and independent consultants, as well as the issue of consultants' conflict of interest, will be addressed in the next section of the paper.

The price dealers charge for services is moving from a cost traditionally embedded in product margins to a fee-for-service arrangement. Movement to a system of *a la carte* pricing allows those farmers who choose not to consume dealers services (for example farmers not using any services, sourcing expertise internally, or those using an independent crop consultant) to pay a price for products that does not include a subsidy directed to other farmers who do use dealer services. As wholesalers and manufacturers participate more actively in the retail sector, dealers must "unbundle" services from products in order to keep their prices competitive. In addition to charging fees for individual services, some dealers are offering comprehensive service packages featuring fall soil testing, manure analysis, planter calibration, seedling counts, weed scouting, pest scouting, tissue testing, and other agronomic practices. Charging service fees is part of the trend toward formalizing services. While farmers are likely to object to paying for previously "free" services, dealers will need to generate revenues in order to invest in equipment and personnel needed to generate and apply detailed information.

Transitions related to service pricing are proceeding in a very uneven fashion across dealerships. Most dealers envision an incremental transition from loss leader (dealership losing money on a service offering is offset by other benefits of providing the service), to cost recovery (break even), to profit maker. Many dealers expressed an experimental approach to service offerings and had explicit plans to evaluate operations and revise their service plans and price schedules after each season.

Dealers are observed to be confronting economic risks by moving strategically to "lock-in" their customer base. Movement toward providing one-stop-shopping and doing more custom application of farm chemicals in combination with growth in information services is serving to increase the level of integration of dealers into on-farm operations. Dealers are attempting to solidify their relations at a time when independent consultants, other dealers, manufacturers representatives, and other prospective advisors are competing for the farmers' attention.

Many dealers are segmenting local farm clientele and concentrating on what was described more than once as the "20 percent of farmers responsible for 80 percent of the sales." This trend is represented by sliding fee scales for custom application services including free service for the largest accounts. Large farmers are being offered "harvest-terms" operating loans (fertilizer, herbicide, seed, and service costs are carried by the dealer until after the harvest at no interest) and variable pricing of farm chemicals (charges for farm chemicals are assessed after the growing season, at which time farmers pay the lowest price at which the particular chemical is sold at any time during the season, regardless of the price at the time they took delivery). Additionally, in some markets, prices for products and services vary according to the size of the farm operation and their relationship with that dealer.

Integration into farming operations and, more directly, fostering farmers' dependency allows dealers to operate under conditions where they have consistent and predictable demand for their products and services. Through information service development and through expansion of custom farm chemical application services, dealers are able to exercise enhanced control over farmers' geographically and temporally uncoordinated requests for production assistance. By taking a more active leadership role in farm and field operations, dealers are able to engage in "just-in-time" management and exercise logistical control over their physical plant, farm chemical, and machinery inventories while utilizing their trained personnel more effectively.

While expansion of dealer services is partly attributable to their strategic interests, the capital intensity, management sophistication, and expectations of local farmers are strong determinants of the extent to which dealers are engaged in service delivery. Leading farmers are exerting substantial pressure on dealers to upgrade their capital resources (e.g., buildings, mixing and loading facilities, inventories, application equipment, automated billing and accounting systems) and their human resources (e.g., custom applicators, agronomists, plant pathologists).

To some extent dealers are observed to be "pushing" their way into farm- and field-level management in order to secure and expand their role in agricultural production systems. Simultaneously, increasingly large, sophisticated, and demanding farmers in competitive markets are "pulling" dealers into positions of greater responsibility.

Independent crop consultants

The independent crop consulting industry is the beneficiary of the same structural, economic, regulatory, demographic, environmental, and technological trends impacting farmers noted above. The independent crop consulting industry is highly diverse. My observations suggest that most practitioners work alone or in very small firms, but larger group practices are increasingly common. Consultants were opening new businesses in most of the study sites. Marketing is done through word of mouth and direct solicitation by consultants. Consultants tailor their services and fees according to negotiations with individual farmers. Independent crop consultants charge more for services than do dealers. The reputation of the individual consultant among local farmers, rather than price of services or the intensity of services (i.e., visits per week), is the key to success.

Consultants in the midwest corn production sites specialize in fertility analysis although some offer some insect and disease scouting services and other advice on topics such as weed management, crop seedling counts, crop rotation, tillage, and cultivation selection. In the Mid-Atlantic site, consultants provide integrated crop management services including soil fertility recommendations and weekly pest scouting services. Farmers growing vegetables and fruits rely on consultants far more frequently than food and feed grain producers. In the cotton production sites independent consultants generally serve as insect scouts. In fact, in those systems, consultants are often referred to as entomologists. It is common practice in these sites for consultants to hire high school age, minimally trained "runners" to scout fields. The entomologist's scouting time based on data provided by runners. This practice allows consultants to cover more acres and keep fees low but likely diminishes the quality of recommendations.

The relationship between time invested per acre by consultants—hence, amount, type and detail of data informing a crop management recommendation—is critical. Several consultants who provide pest scouting services expressed that they could not charge higher fees without losing customers, yet could not afford to gather as much information as they would like to in order to understand the ecological interactions in each of their clients' fields. In the Mississippi higher competition site there is a pilot study to evaluate the economic and agronomic performance of "as needed" scouting services. An "as needed" approach is one in which consultants do not maintain their standard once a week field visit schedule but instead go to fields whenever biophysical conditions warrant a visit, and then bill farmers accordingly. Reportedly, this approach will result in higher per-acre consulting fees as well as higher per-acre net farm income. Despite the merits of alternative approaches to development and application of information to input use decisions constraints on farmers willingness and/or ability to pay for informational services is serving to retard growth in the quality of consultants' services.

Competition between independent crop consultants is not a pronounced feature of any of the production systems visited. Many consultants were not accepting new clients and were cutting back the number of acres on which they provide services. Independent consultants were widely regarded as highly knowledgeable. With few exceptions, farmers employing consultants were impressed with their diligence, flexibility, and level of personal investment in the success of the crop. The attributes farmers valued most highly in their independent consultants were scientific knowledge and practical experience, willingness to devote sufficient time in individual fields to do a thorough job, and the fact that the consultant worked directly for them rather than an agrichemical company.

Despite great opportunities for growth in the independent consulting industry there are significant obstacles. The analysis below focuses on two related issues central to the success of an independent consulting industry, competition in service markets from dealers, and mergers with dealers.

Competition from dealers

Dealers have a number of advantages in service markets over independent consultants. Advantages include a larger resource (capital) base, ability to capitalize on economies of scale applied to things such as lab facilities, ability to cross-subsidize services with product revenues, and more direct relations with input manufacturers. Also, dealers enjoy a logistical advantage in farming systems as farmers can conceivably conduct one-stop-shopping through their dealers—obtain products and production recommendations and in some cases marketing services—but not through their independent crop consultant. It will be clearly be difficult for independent crop consultants to compete with dealers prepared to offer discounted products, free custom application, interest-free financing, and below cost fertility and pest management services. Competition from dealers becomes especially problematic when services are provided by trained personnel who make consistent field visits, have access to substantial technical resources, provide written recommendations, keep complete records, and "warranty" their recommendations. If independent consultants cannot convincingly persuade farmers that their informational services are more valuable than those offered by dealers, dealers will dominate local markets. Independent consultants' comparative advantage over dealers stems directly from their independence, obviously. As independent consultants tend to specialize in particular crops and field services, and many of them have been intensively engaged in service provision for many years, they are likely to have advantages over dealers just entering into service markets. While the experiential advantage may fade over time, the issue of bias is likely to have lasting currency in the eyes of some farmers.

Independent crop consultants and dealer linkages

In the higher competition sites within each of the three states independent consultants are entering into alliances with dealers. Under these arrangements farmers will be served by independent consultants working under contract or negotiated agreement with a local dealer. While more dealers are engaging in service delivery and the number of independent consultants is increasing, alliances between independent consultants and dealers will result in a decrease in the number of discrete information outlets. More important than the raw number of information outlets, the potential decrease in the number of non-agrichemical industry affiliated outlets would appear to be a topic of concern.

The contractual mechanisms supporting mergers between dealers and independent consultants include sub-contracts, guaranteed referrals, and exclusive service contracts. For example, in the higher competition Midwest site a local dealer was paying 75 percent of a local independent consultant's fee on customers' acreage. In such cases, dealers benefit from outsourcing services as they avoid disrupting in-house activities and the process of reallocation of human capital. Additionally, dealers benefit by maintaining flexibility, delaying or avoiding certification and licensing requirements, and positioning themselves to respond to claims of conflict of interest. It is assumed that consultants entered into these arrangements after having considered the prospect of competing head-to-head with dealers. Some consultants believe the advantages held by dealers as discussed above make it difficult for an independent consultant to compete successfully.

These linkages between consultants and the agrichemical industry make the concept of independence difficult to define. While independence is always relative, a subcontract with an agrichemical supplier is difficult to reconcile with the label "independent." Presumably, independent crop consultants providing services through a dealer will be subject to pressures not experienced by independent crop consultants working one-on-one with farmers. As one independent consultant who frowned upon mergers with dealers said, "If the consultant doesn't make recommendations that please the dealer, how long do you think the consultant will be working for them?" The traditional client of independent consultants has been the farmer. Due to factors discussed in this paper, this relationship is changing to incorporate dealers. As agricultural information markets are privatized and commercialized, the success and autonomy of the independent consulting industry would appear to be an important check on formation of an industrial "information monopoly."

QUALITY OF INFORMATION AND CONFLICT OF INTEREST

The variety of fertility and pest management services available to farmers is rising, and the marketing effort associated with these services is intensifying. Taken at face value these developments bode well for the environment. A more complex aspect of this phenomenon involves assessment of the quality of information provided by consultants. As dealers and independent consultants emerge as powerful determinants of the sophistication of farm chemical management, the quality of input recommendations provided by private firms becomes increasingly important. Economic considerations such as data collection and analysis costs impact the quality of recommendations. Scientific considerations such as limited understanding of pest ecology limit predictive capability. Human capital constraints limit the quality of recommendations as consultants lacking specific training likely make avoidable errors. Additionally, ethical considerations and the presence of contradictory incentives effect information quality.

Bias associated with a conflict of interest is a serious concern as reflected in the strategic activities of national organizations representing crop advisors and community-level observations. While the integrity of individuals is at issue, the focus here is on identifying structural problems in information markets, specifically relationships between the advising function and the product sales function in agriculture. Three situations presenting a conflict of interest were observed in the case study site visits. First, situations in which dealers sell farm chemicals and advise producers on extent and intensity of use of those products is a source of concern. Second, independent consultants are receiving informal compensation in the form of gifts, free travel, and other benefits from farm chemical manufacturers' representatives, distributors, and dealers. Third, alliances, mergers, subcontracting arrangements, and exclusive referral services serve to blur the distinction between agrichemical dealers and independent crop consultants. Each of these arrangements were identified by interview subjects as problematic and as leading to unnecessary input use.

In specific communities, dealers and independent consultants are perceived by some farmers and other people involved in agricultural production systems as motivated by interests other than the farm client's. Dealers themselves, as well as many other interview subjects, expressed that "dealers oversold fertilizer and pesticides in the past." Some independent consultants, farmers and Extension personnel perceive specific dealers as continuing to recommend more inputs per acre than warranted. Independent consultants, specifically entomologists in the higher competition Mississippi Delta state, are perceived as providing biased recommendations in return for gifts and travel offered by agrichemical companies. Despite widespread acknowledgment that some dealers and consultants provide biased information, the vast majority of farmers expressed faith in their information providers. Farmers also expressed confidence in their own ability to recognize bogus recommendations.

Conflict of interest implies bias derives from dealers and consultants benefiting from incentives and kick-backs from chemical manufacturers, distributors, or retailers. However, there is a separate and more subtle form of bias inherent in consulting markets. Farm advisors have an incentive to err on the side of caution, meaning interpret pest treatment thresholds with an eye on minimizing exposure to blame. As one consultant expressed, "It is easier to prove you should have sprayed than you did not need to spray." Farmers, Extension, dealers and independent crop consultants expressed that there are incentives to apply inputs when the data informing such a decision are inconclusive or open to interpretation. Managing a crop to maximize a farmer's net income is far more risky for a consultant than is managing for maximum yield and highest quality. Recommendation of an additional fertilizer or pesticide application is preferred by consultants to risking crop yield and/or quality loss. This form of bias stems from consultant's interest in insuring their reputation, their most important asset.

The data required to quantitatively determine the extent to which conflict of interest and/or an "insurance" mentality results in over-application of inputs are not available and are likely to be extremely difficult to develop. These case study data indicate that the problem is significant. Apart from the environmental and farm productivity impacts, the perception that some consultants provide biased information must be regarded as a serious impediment to development of a crop consulting industry.

CONCLUSION

This paper has described the emerging structure and patterns of relations that define crop production information markets. These data support the hypothesis that competition in local markets and processes of restructuring of input supply chains are motivating firms to develop their farm chemical management service capabilities. It is not clear from these data that expansion of commercial consulting markets translates into environmental protection benefits. However, political and economic processes are serving to fuel professional certification programs, investment in equipment and human capital, development of formal agronomic service programs, and heightened attention to detail among crop advisors. These behavioral or "process-oriented" measures of information quality indicate potential material improvement in the efficiency of farm chemical use.

Local agrichemical dealers and independent crop consultants significantly determine the quantity and quality of information applied in making fertilizer and pesticide management decisions on many farms. Due to growth in private sector services and redirection of public sector investment, Extension occupies a small and diminishing role with respect to direct provision of crop management recommendations. It is clear that private consultants will be increasingly integrated into on-farm management as economic, social, technological, institutional, regulatory and agroecological pressures continue. The strengthening of linkages between farmers and private sector consultants is consistent with first, the growth in demand for and value in highly applied information; second, a diminished governmental presence in agriculture; and third, increasing levels of integration and coordination within an industrializing agro-food system.

In an era in which information is recognized as an environmental management tool and a source of economic competitive advantage, commercialization of information is regarded as highly significant. As agricultural Extension becomes a less important provider of field-level recommendations and fee-based consulting markets expand, researchers and policy makers must respond to questions surrounding structural change in agriculture and the environmental performance of cropping systems. This study indicates that larger farms featuring more sophisticated management located in more intensive production regions will have advantages in obtaining informational inputs from commercial sources. In theory, these farms will exploit their competitive advantage advancing the existing bimodal structure of agriculture. This conclusion is based on the observations that the distribution of private sector information infrastructure is geographically heterogeneous and consultants' incentives for working with specific segments of the farm sector are biased toward larger farms. To the extent that political support exists for maintenance of large numbers of small and medium-sized farms and for farms operating in areas featuring lower levels of private sector information infrastructure, there is an argument to be made for (re)dedication of public sector resources.

Public scrutiny focused on environmental, ecological, and economic costs of farm chemical pollution indicate that there will likely be demand for oversight over the development of farm chemical recommendations. The salience of environmental and consumer groups' claims applied to farm chemicals as well as those of farmers concerned about being "oversold" products suggest that the issue of conflict of interest will gather momentum rather than dissipate. Under such circumstance, government intervention through Extension and other agencies, certification and education programs, and cooperative activity among farmers represent potentially positive approaches to promoting availability of high quality information.

Maintenance of a competitive structure in local consulting markets is a key to promoting both innovation and information quality. Therefore, rapid concentration within the agrichemical supply industry and the blurring of the independence of some consultants due to increasingly close relations with agrichemical interests is viewed as problematic. These developments directly limit farmers' opportunities for validating farm chemical management recommendations. In the absence of a competitive market and locally active Extension Service, the mechanism supporting development of high quality information is not clear. As privatization of agricultural production information proceeds, research will play an important role in defining how to effectively integrate market processes and public institutional involvement.

NOTES

[1] Other types of production consultants include input manufacturers' representatives, seed company employees, fieldmen working for a processing company, machinery and computer software service providers, farm managers inside and outside of lending institutions, employees of governmental agencies such as Extension and the Natural Resources Conservation Service, and technical specialists working for resource conservation organizations such as The Nature Conservancy.

[2] Private funding of agricultural research surpassed public investment sometime during the 1980s. While public funds remain a large and vital component of total investment, the transition from public to private leadership in financing of agricultural research is vitally important in its own right, as well as being symbolic of the shifting role of the state in agriculture.

[3] For example, sugar beet growers in the Red River Valley who have adopted precision farming techniques to manage fertility have "invariably gone to higher phosphorous application rates" (Freeberg 1996). Further, crop consultants, many of whom are independent crop consultants, lobbied USDA in connection with performance standards governing cost sharing of integrated crop management (ICM) under the Water Quality Incentives Program (WQIP). The original set of rules considered for the program called for a 20 percent reduction in agrichemical usage as a program eligibility requirement. Consultants argued that the program should be designed for *informed use* of agrichemicals, not necessarily a *reduction in use*. The absolute reduction requirement was eliminated from the program.

[4] What is now the Cooperative Extension Service was originally a private sector venture. Seaman Knapp, acknowledged as the first extension agent, was originally hired by railroad interests to bring scientific agriculture to the countryside in order to boost freight volume (McConnell, 1969).

[5] The specific plans were ICM plans (SP-53) written in conjunction with the WQIP as directed in the 1990 Farm Bill.

REFERENCES

Bennett, C.F. 1994. Rationale for public funding of sustainable agriculture extension programs. Paper presented at American Evaluation Association Annual Meeting. Vancouver, BC, Canada. November 1-5.

Bocher, L.W. 1990. Corn belt consultants gaining ground. Agrichemical Age. March, pp. 6-8.

Bonanno, A. 1990. Agrarian Policies and Agricultural Systems. Boulder: Westview Press.

Busch, L., W. Lacey, J. Burkhardt, and L. Lacey. 1991. Plants, Power, and Profit. Oxford: Basil Blackwell.

Buttel, F.H. 1991. The restructuring of the American public agricultural research and technology transfer system: Implications for agricultural Extension. In Agricultural Extension: Worldwide Institutional Evolution and Forces for Change W. Rivera and D. Gustafson (eds.). Amsterdam: Elsevier

Buttel, F.H. 1992. Environmentalization: Origins, processes, and implications for rural social change. Rural Sociology 57:1-27

Dysert, J. 1994. Green movement pushes to make crop advice illegal. Custom Applicator, Oct.

ERS (USDA/Economic Research Service). 1995. Agricultural research and development: Public and private investments under alternative markets and institutions. Staff Paper 9517. Washington, DC: USDA.

Ford, S.A., and E.M. Babb. 1989. Farmer sources and uses of information. Agribusiness, 5(5):645-476.

Freeberg, M. 1996. Paper presented at the annual meeting of the National Alliance of Crop Consultants. Orlando, FL. January 24-27, 1996.

Fuglie, K.O., and D.J. Bosch. 1995. Economic and environmental implications of soil nitrogen testing: A switching regression analysis. American Journal of Agricultural Economics. 77(4):891-900.

Gannon, E.M., and R.G. Ginder. 1992. Evaluating future strategies for Iowa farmer-owned cooperatives in supplying agricultural products and services: An assessment of integrated crop management services. Leopold Center for Sustainable Agriculture, Ames, IA.

Ginder, R. 1992. The future role of farm input suppliers in the sustainable agriculture movement. Staff paper #241, Iowa State University.

Goe, R.W. 1986. U.S. agriculture in an information society: Rural Sociological Research. Rural Sociology 6(2):96-101.

Goodman, D., B. Sorj, and J. Wilkinson. 1987. From Farming to Biotechnology. Oxford: Basil Blackwell.

Hebert,T. 1995. Paper presented at the annual meeting of the Soil and Water Conservation Society. Des Moines, IA.

Hercus, J.M. 1991. The commercialization of government agricultural extension services in New Zealand. In Agricultural Extension: Worldwide Institutional Evolution and Forces for Change, W.M.Rivera and D.J. Gustafson (eds.). Amsterdam: Elsevier.

Hoffman W.L. 1993. Stemming the flow: Agrichemical dealers and pollution prevention. Environmental Working Group, Washington, DC.

Keeney, D. 1991. Science with stewardship: The role of input dealers in sustainable agriculture. Leopold Center for Sustainable Agriculture newsletter, 3:3.

Keystone Center. 1995. The Keystone national policy dialogue on agricultural management systems and the environment. Keystone, CO. May.

Lambert, H. 1995. Paper presented at the Privatization of Technology and Information Transfer in U.S. Agriculture Workshop. Madison, WI, October 25-26.

Lambur, M., R. Kazmierczak, Jr., and E. Rajotte. 1989. Analysis of private consulting firms in integrated pest management. Bulletin of the Entomological Society of America, Spring:5-11.

McConnell, G. 1969. The Decline of Agrarian Democracy. New York: Atheneum.

NRC (National Research Council). 1993. Soil and Water Quality: An Agenda for Agriculture. Board on Agriculture. Washington, DC: National Academy Press.

NRC (National Research Council). 1992. Global Environmental Change. Washington, DC: National Academy Press.

NRC (National Research Council). 1989. Alternative Agriculture. Washington, DC: National Academy Press.

Nowlin, B. 1993. NAICC report. AgConsultant. Fall: 13.

OTA (Office of Technology Assessment, US Congress). 1990. Agricultural research and technology transfer policies for the 1990's. A Special Report of OTA's Assessment of Emerging Agricultural Technology - Issues for the 1990s. OTA-F-448. Washington, DC.

Peitscher, A. 1991. Will there be enough to go around? Solutions, Nov./Dec.: 12-14.

Rivera, W.M., and D.J. Gustafson (eds.) 1991. Agricultural Extension: Worldwide Institutional Evolution and Forces for Change. Amsterdam: Elsevier.

Rivera, W.M., and J.W. Cary. (in press) "Privatizing" agricultural extension: Institutional changes in funding and delivery of agricultural extension. In Improving Agricultural Extension: A Reference Manual, B.E. Swanson (ed.). Rome: Food and Agriculture Organization.

Robert, P., R. Rust, W. Larson (eds.). 1995. Site-specific Management for Agricultural Systems. Madison, WI: American Society of Agronomy, Crop Science Society of America, and Soil Science Society of America.

Robert, P., R. Rust, and W. Larson (eds.). 1993. Soil Specific Crop Management. American Society of Agronomy, Crop Science Society of America, and Soil Science Society of America. Madison, WI.

Schnitkey, G., M. Batte, E. Jones, and J. Botomogno. 1992. Information preferences of Ohio commercial farmers: Implications for Extension. American Journal of Agricultural Economics, May: 486-496.

Sine, C. 1990. The emphasis shift to services. Farm Chemicals. Summer 153(13):90.

Smith, V.H. 1996. Causes and consequences of agricultural research and development policies in four developed countries: Implications for research. Paper presented at the NC208 Symposium, New Issues in Research on Agricultural Research, Washington, DC March 29-30, 1996.

Swinton, S., and King. 1994. The value of information in a dynamic setting: Case of weed control. American Journal of Agricultural Economics, Vol. 76(1):36-46

Vandeman, A., J. Fernadez-Cornejo, S. Jans, and B. Lin. 1994. Adoption of integrated pest management in U.S. agriculture. Resources and Technology Division, ERS/USDA, Agricultural Information Bulletin No. 707.

Watts, S. 1995. Paper presented at the Privatization of Technology and Information Transfer in U.S. Agriculture Workshop. Madison, WI, October 25-26.

Wolf, S. 1996. Privatization of crop production service markets: Spatial variation and policy implications. Unpublished Ph.D. thesis, University of Wisconsin-Madison, Institute for Environmental Studies.

Wolf, S. 1995. Cropping systems and conservation policy: The roles of agrichemical dealers and independent crop consultants. Journal of Soil and Water Conservation 50(3):263-70.

Wolf, S., and F.H. Buttel. (1996). The political economy of precision farming. American Journal of Agricultural Economics. Vol. 78 (December).

Wolf, S., and S. Wood. (forthcoming).Precision farming: Environmental legitimation, commodification of information, and industrial coordination. Rural Sociology.

Wolf, S., and P. Nowak. 1995. The status of information-based agrichemical management services in Wisconsin's agrichemical supply industry. In Site-Specific Management for Agricultural Systems, P. Robert, R. Rust, and W. Larson (eds.) Madison, WI: American Society of Agronomy, Crop Science Society of America, and Soil Science Society of America.

Section Three

INTERNATIONAL CASES

10

Issues in Public and Private Technology Transfer: The Cases of Australia and New Zealand

John Cary

The reconsideration of the public delivery and funding of extension may represent responses to political fashion and budgetary circumstances. It also represents a significant recognition of the changed nature of agriculture in developed economies and changed circumstances in the wider economies of many nations. Many developed countries are reducing their commitment to publicly funded and publicly delivered extension. In some cases this is for political and fiscal reasons. In most cases it is because agriculture has lost its 'special' characteristics or, in Paarlberg's (1978) words, 'its uniqueness.' Agriculture has become more commercialized and little differentiated from other commercial businesses. In the face of declining public financial support, there is a paradox for publicly delivered technology transfer: more sophisticated and more specialized agriculture requires more specialized and much more individually tailored information to be provided to individual farmers operating complex systems of farm management. Such demands are likely to require more extensive one-on-one dealings between extension advisers and farmers.

Reconsideration of public extension has occurred within a wider context of reviews of government support for agriculture as well as government support for other industries and other public activity in the economies of many countries. Australia and New Zealand have not experienced these changes in their agricul-

The helpful comments and suggestions of Roger Wilkinson of Landcare Research New Zealand Ltd., Nigel McGuckian, Rendell McGuckian Bendigo, and Geoff Mavromatis of Agriculture New Zealand are gratefully acknowledged.

ISBN 1-57444-104-3/98/$0.00/$.50

tural sectors as recently as many other developed economies. In both Australia and New Zealand, the agricultural sectors have a relatively long history of producing and trading in more or less 'free' domestic markets and, particularly, of selling a large proportion of their agricultural outputs in export markets over which they exert little control. These two countries also have had relatively little direct government intervention in farmers' production decisions, and thus no history of seeking cross compliance initiatives at the farm level. As a consequence, New Zealand and Australia have had a head start over many other developed countries in thinking about, and implementing, change in the public funding and delivery of extension. For other countries, as the political influence and the uniqueness of agriculture declines, there may be lessons to be learned from the experience of New Zealand and Australia.

The New Zealand government has fully privatized the former Ministry of Agriculture and Fisheries advisory service. In Australia, a range of less radical initiatives to restructure the delivery of agricultural extension services have been undertaken or are being considered by major State agencies. The various options that have been considered or implemented in Australia and New Zealand can be categorized in terms of different combinations of exclusively public, partially public, or private funding; and exclusively public, partially public, or private delivery.

In determining the appropriateness of extension delivery by the state or private agents a number of issues need to be explored. First, the differing contexts for evolutionary and more significant structural change in the United States, Australia, and New Zealand are surveyed. Second, privatization and commercialization are differentiated on the basis of agency ownership. Broader examples of 'privatization' are considered within a framework of sources of funds provision and modes of service delivery. Third, new structures for extension delivery in New Zealand and Australia are presented. Fourth, the rationales underlying private and institutional provision of technology transfer are explored. Fifth, new modes and new challenges for the delivery of public good environmental extension are identified. Finally, there is a consideration of procedural and transitional difficulties in moving from one form of delivery to another.

STRUCTURAL CHANGES VERSUS FASHION CHANGES

In teasing out which territory is best served by public enterprise or private enterprise it is a difficult task to separate desirable, evolutionary structural change, social or political fashion change, or political opportunism. Each of these forces contributes to the existing balance of public and privately delivered extension.

Historical factors are also important in determining the culture for public or private delivery. The United States has had a long and extensive experience of publicly-funded extension delivered through the land grant colleges. This situation was preceded, between 1900 and the passage of the Smith-Lever Act in 1914, by active extension programs delivered by private railroad companies concerned to 'promote better agricultural methods' in the interest of 'increased traffic of farm supplies and farm products' (Kloppenburg 1988). Kloppenburg presents evidence that the public funding for extension that came with the passage of the Smith-Lever Act was largely engineered by lobbying by private transportation firms, manufacturing companies, and banks.

Acknowledging the differing histories of 'white' land settlement is also informative in understanding the evolution of extension delivery systems whose rationales are now being rethought. In this vein, it is useful to compare Australia and the United States. Australia's early history and subsequent institutions were dominated by its being a colony of Britain. There was a long period of dominant administrative influence associated with British 'rent a government', much the same as has existed until recently in Hong Kong. With the exception of the initial period of relatively uncontrolled pastoral tenancy, early land settlement was strongly bureaucratically influenced by, albeit not always in wise response to, local political pressures. Between 1861 and 1869, the colonies of New South Wales, Victoria, Queensland, and South Australia introduced land selection acts which made land available to settlers at low cost, in a similar way but not as generously as the U.S. Homestead Act of 1862. The United States reviewed its relationships with Britain with more finality with the War of Independence. In the United States, land settlement, while determined by government legislation, tended to have less government intervention in the selection and settlement process. More importantly, in Australia, failed private attempts to introduce irrigation on extensive areas of low-rainfall land resulted in large-scale government irrigation and land settlement programs from the 1880s until the 1970s. Often these projects were fraught with subsequent economic and ecological difficulties for farmer settlers (Barr and Cary 1992). Much of the momentum for early government extension in Australia was predicated on overcoming problems associated with such land settlements. The United States extension system, linked to the land grant colleges, has had a much more educative focus than publicly funded, problem-centered extension in Australia.

New Zealand, while having a history of violent land disputes with indigenous inhabitants in common with the United States and Australia, did not have the settlement difficulties experienced in Australia. With a climate and pastoral growing conditions much more similar to the United Kingdom, New Zealand adopted an extension system with an advisory focus, similar to the Agricultural Development and Advisory Service in England and Wales. An important difference between Australia and New Zealand, with respect to the ease in which changes can be made in government intervention, is that New Zealand has a single unicameral parliament. Australia has a federal system with two parliamentary houses; as well, most states have bicameral parliaments.

WHAT IS PRIVATIZATION?

In the broad sense, extension potentially involves a complex of public and private activities. Thus, multiple arrangements exist for the funding and the delivery of agricultural extension. Tables 1 and 2 illustrate combinations of public and private payment for, and public and private delivery of, agricultural extension services for New Zealand and Australia respectively.

Privatization is often used in a fairly loose way when applied to restructuring of the delivery of funding and agricultural extension services (Rivera and Cary 1995). In its pure sense, privatization implies a full transfer of ownership (usually by way of sale) from government to a private entity, with that entity then meeting all costs and receiving any profits.

In the case of commercialization the government retains ownership but the agency has a separate board, an accounting and reporting system independent from the government, and a remit to 'make a profit' or to be more or less self-funding. Commercialization implies charging for services while retaining the agency in public ownership, and agency personnel who are not government employees. Other possibilities for governments to reduce their direct delivery (and, often, their funding) of extension include shifting public sector delivery services to private sector by contacting, or 'outsourcing', delivery of the service while maintaining oversight and basic funding. A common means of reducing government outlays is to pursue cost recovery measures and fee assessment to pay for the services provided by a government agency. These forms of commercialization are not always independent: a mix of systems may be present; or one may lead to another as an institution undergoes change.

THE NEW ZEALAND EXPERIENCE

Traditionally, extension in New Zealand was delivered by a department of government, rather than through a land grant college system as in the United States. The New Zealand Ministry of Agriculture and Fisheries advisory service

Table 1. Institutional Changes in Funding and Delivering Agricultural Extension in New Zealand[a]

		WHO PAYS	
		Public	Private
WHO DELIVERS	Public	Ministry of Agriculture (production and monitoring: public good activities only) Regional Councils (regional natural resource conservation: funded by regional rating levies) Ministry of Conservation (National estate, public land)	In the transitional 'commercialization' situation prior to full privatization: Fee-based or contract-based services paid by farmers
	Private	Technology transfer projects (currently 9) delivered by Agriculture New Zealand, funded by the Foundation for Research, Science and Technology; Stewardship monitoring farmer groups delivered by Agriculture NZ, funded by MAF.; and monitor farms program funded by the Meat Research & Development Corporation.	Extension delivery by private agricultural and management consultants, or professionals employed by input suppliers. Agriculture New Zealand engages in larger commercial program sponsored by agribusiness firms.

[a] Based on a typology developed by D. Neilson (1993 unpubl.).

Table 2. Institutional Changes in Funding and Delivering Agricultural Extension in Australia

		WHO PAYS	
		Public	Private
WHO DELIVERS	Public	In each State : Departments of Agriculture and Natural Resources (production, and private land resources) Departments of Natural Resource Conservation (public and private land)	Some corporate sponsored environmental programs in Australia)
	Private	Outsourcing of production extension delivery in Victoria. Proposed outsourcing of environmental extension delivery in Victoria. Government rural adjustment authorities subsidise preparation of business plans for farmers by business consultants.	Extension delivery by private agricultural and management consultants, or professionals employed by input suppliers. Farm Management 500 in South Eastern Australia.

(MAF) has been the first publicly funded and publicly delivered extension service to be fully privatized from government ownership.

The sequence of the transition is of interest (Table 3). In 1985, the New Zealand government introduced a user-pays philosophy. Subsequently, in 1986, the government required the extension division of MAF to charge for advisory services provided to farmers. As well, a national accounting and charging regime was established for services the MAF field staff provided for central government (such as reporting district seasonal conditions, provision of advice and assistance for local disaster relief, etc.). In 1987, the Advisory Services Division was amalgamated with MAF's Agriculture Research Division to form MAF Technology. In

Table 3. Privatization Transition of Agriculture New Zealand Ltd

Year	Activity
1985	NZ government introduces user-pays philosophy
1986	MAF Extension charges for advisory services
1987	Advisory Services Division amalgamated with Agriculture Research Division as MAF Technology.
1990	Advisory activities restyled as Management Consultancy Service
	Management Consultancy Service established as separate business
1992	State Owned Enterprise proposal aborted
	MAF Consulting renamed Agriculture New Zealand, with a requirement to be self funding
1994	Agriculture New Zealand self-funding
	Staff purchase offer for the business (preferential right) rejected by NZ government
1995	Agriculture New Zealand sold to Wrightson Ltd.

1990, the advisory activities were restyled as the Management Consultancy service and established as a separate business.

The service was fully commercialized in 1992. It had been intended that the Management Consultancy Service of the Ministry of Agriculture and Fisheries become a state owned enterprise (SOE). For various reasons, mainly related to staff acceptance and perceptions of employment and redundancy implications, the SOE proposal was not implemented. Instead, commercialization of the organization was pursued (Ritchie undated). MAF Consulting was renamed Agriculture New Zealand, with a requirement to be self funding. Government funding was to be provided only for district intelligence and disaster services provided, under contract, to the government. The agency depended for its annual budget on consulting fees received from farmers and contractual arrangements with government. Initially financial short falls, where fees charged did not meet administrative and salary outlays, were met by the government. MAF employees gave up a number of public employment benefits and received commissions for consulting work undertaken.

In 1994 the number of consultants employed in Agriculture New Zealand was about half of the number employed prior to commercialization; at the time of full privatization the number had fallen to less than a third. There was a significant downsizing from 450 advisers in the mid 1980s to about 100 consultants in 1995 (Ritchie undated). Some advisers retired or departed voluntarily; others established private consulting businesses. The consequence of the changes in New Zealand was an increase in fee-for-service consulting (the number of private farm and horticultural consultants approximately doubled), with the traditional public 'advisory' extension no longer existing.

By 1994, the former MAF consulting service was self-funding. The government then sought to fully privatize the service (by offering it for sale). Having independently established a value for the goodwill, the rights for contracting services to government, and certain other assets, the government first offered the 'business' to MAF employees. There was an apparent large gulf between the government's valuation and the employee's offer price for the ongoing business.[1] The agency was subsequently sold to Wrightson Ltd., New Zealand's largest rural input supplier and selling agency, which is publicly listed on the New Zealand stock exchange. The Wrightson Group consists of a central company involved in rural real estate, merchandising livestock, trading, deer, wool, as well as a number of subsidiaries. The subsidiaries include Wrightson Finance Ltd., Wrightson Seed Ltd., Wright Bloodstock Ltd., Agri-feeds Ltd., Agriculture New Zealand Ltd., and a specialized wool and hide marketing company. Since privatization, an active recruiting campaign has increased the number of consultants employed to just over 100, which was the size of the professional staff in 1994.

There has been no formal assessment of the New Zealand changes. The earlier commercialization experience was perceived to have had a positive effect of moving extension staff 'beyond the farm gate' to an involvement in the broader production-processing-transporting-marketing chain. There was a shift to a client orientation, and a concern to produce results rather than simply to engage in activities (Hercus 1991). However, Walker (1993) suggested that there appeared to be less interaction between organizations, reduced feedback from farmers to science providers, and more limited information distribution, particularly to less well-off and poorer performing farmers. While, in most cases, the changes seem to have been readily accepted, there remains concern over the effective transfer of scientific findings into agriculture.

Traditional technology transfer extension is now largely confined to agricultural commodity boards. In New Zealand, extension services to the dairy industry for many years have been financed by the dairy industry and delivered by the dairy board consulting service using discussion groups. However, Agriculture New Zealand has formed a national network, known as Farmwise, to supply commercialized services to the dairy industry. Agriculture New Zealand engages in some specific 'public good' technology transfer projects on a contact basis to commodity research agencies and the National Foundation for Research, Science and Technology. The foundation, in evaluating proposals for research funding, places strong emphasis on technology transfer. Much of this technology transfer is to user groups, at a broader level than traditional extension.

THE AUSTRALIAN EXPERIENCE

In each Australian state the major provider of publicly funded extension services is the respective state Department of Agriculture or Department of Primary Industry. In the last decade there have been significant changes in the nature of extension programs and the level of public resources devoted to extension in Australia. Within the last few years five of Australia's six states have conducted reviews of extension service delivery provided by their Departments of Agriculture. These reviews have generally recommended a refocusing of extension activity and reduction or withdrawal of government-provided advisory services of a direct or face-to-face nature. Despite the contracting resources available for government extension services, in some states there is a high and apparently increased level of contact between government-employed extension personnel and farmer clients (Cary and Wilkinson 1995; Harrison Market Research 1991). The high rate of farmer contact reflects a farming community which is much smaller than a decade earlier and more active in its search for information. As well, in most states there has been increased use by farmers of private or commercial advisers.

A sense of what farmers in Australia seek from government extension agencies can be gleaned from a survey of a sample of 426 Victorian farmers undertaken in 1992. The government agricultural extension agency in Victoria predominantly was seen as a supplier of technical information (Table 4). There were fewer demands for marketing or environmental information, but the demands for servicing environmental and landcare groups were increasing (Cary and Wilkinson 1994). Leaders of government extension agencies often espouse a greater emphasis on marketing and management aspects of farming. The strength of public extension has traditionally been transfer of, or advice about, technology. Farmers in Australia do not appear to have a strong demand for marketing and management extension from public agencies. This may be because government agencies are not perceived to be highly competent in this area, or it may be perceived as the domain of the private sector. Indeed, to give marketing and management advice relevant to the circumstances of an individual firm implies a private good. The alternative is to provide more general extension education about marketing and

Table 4. Most Important Information Needs Which Farmers Considered Should Be Provided by Agriculture Victoria: Unprompted (multiple) Responses (N=415)

Information	Per cent of responses	Per cent of farmers
Technical	28	37
Management and financial	9	12
Marketing	6	8
Environmental and landcare	5	7
Laboratory and testing services	11	14
Whatever needed at the time	16	21
Latest up-to-date information	15	20
Other	10	15
Total	100	134

Source: Cary and Wilkinson (1995)

management; but this is unlikely to have high credibility or usefulness if delivered by extension officers without 'real world' experience.

Proposed institutional changes in extension service delivery include service reduction, commercialization by charging fees for services, and contract delivery by private agents. A review of the delivery of government extension services in the Australian state of Victoria determined that for government-provided services conferring essentially private benefits to individuals it was more desirable and more efficient for private advisers to deliver such services, rather than engaging in cost recovery by government fee charging (Watson et al. 1992). However, because of the complexities of extension service delivery and differing levels of commercial development of different agricultural sectors the review recommended that some forms of publicly funded extension should continue. An important reason for involvement by government followed from the riskiness of Australian agriculture. Those rural industries subject to large and extended fluctuations in income associated with fluctuations in production levels or product prices often cannot sustain viable private advisory and consulting services. Other important reasons for involvement by government were the 'public good' aspects of information, the importance of externalities associated with land resource restoration, and the need for public protection and education related to food quality and plant and animal diseases.

During recent years most new extension staff positions in Agriculture Victoria have been funded as projects with 'initiative' funding from farm industry research corporations or as part of a special program of government. To provide an alternative framework for farm industry research organizations to take a greater responsibility for technology transfer, the Victorian government has proposed 'outsourcing' as a means for delivery of future extension programs. Outsourcing means that the government extension agency will retain a core pool of extension project staff and 'buy in' private sector professional services with skills that the agency considers unnecessary to maintain. Agricultural consultants and contract staff will be employed to help deliver services in specific projects funded by rural industry and the federal government. Such projects are likely to be broad and industry-wide, and not tailored to individual farm circumstances.

Outsourcing implies government-provided funding and privately-delivered services. In fact, one of the attractions of outsourcing for governments is that funding partnerships can be formed between government and industry, or government and rural research and development organizations to jointly fund technology transfer efforts. In Australia in the past it has proven difficult to get such organizations to agree to contribute to government delivered extension services, because such contributions are seen as allowing the government to reduce its funding proportionately.

A brief comment on two other Australian states will give some flavor of other initiatives that have occurred. Tasmania, the smallest Australian state, was the first to seek to 'commercialize' extension services. Tasmanian extension emphasized the delivery of technical and management advice. In the 1980s, the Tasmanian government commenced charging fees for agricultural advisory services. The experiment did not meet with widespread success. Only a few extension officers generated sufficient advisory 'business' to ensure a viable commercial outcome. Often, officers found ways to circumvent charging farmer clients for their services. And there was a substantial decline in demand for extension services for which a fee was charged.

In Tasmania the most obvious reason for the commercial failure of fee for service was because the advisers didn't have a financial interest in generating fees. Generally, they did not have skills which could be marketed to potential farmer clients. Some extension officers did not have the commercial 'mind set' to send an invoice after delivering a service. It is likely farmers did not sufficiently value the type of advice on offer, or the professional quality of the advice was not considered worth the price.

In South Australia, agricultural extension and advisory services have also been commercialized and focused on 'marketing' new agricultural initiatives. At the same time, a very large reduction in government funding for extension and advisory services was instituted. A move to fee charging would seem to be a consequence of the funding reduction rather than vice versa. Given the obstacles of transforming advisers on government salaries to commercial consultants and the transitional difficulties that need to be overcome, it is unlikely that the South Australian approach will be successful. Generally, it makes little sense to institute commercial fees for government services which are able to be delivered efficiently by commercial providers, except as a transitional measure.

In most countries, private sector companies are important contributors to technology transfer and the advancement of agricultural development mainly through contract arrangements with farmers. In Victoria, the private sector has come to be acknowledged as a major information provider to large and small scale crop farmers. (Cary and Wilkinson 1992).

In Australia a commodity 'tax' is levied on most rural production; this revenue is matched by the federal government (on a dollar for dollar basis, although there are current recommendations to reduce the government contribution). These levies provide the financial basis of statutory commodity research and development corporations which allocate funds for both fundamental and applied agricultural research. The charters for these organizations, as for New Zealand's Foundation for Research, Science and Technology, include technology transfer of new research findings. To date, relatively small amounts of resources have been devoted to this purpose. However, the structure exists to provide private funding (via the commodity 'tax') for both publicly and privately delivered technology transfer and possibly other extension services.

The changing nature of federal government funding for extension has played a considerable part in changes to delivery systems at the state level. During the period 1966 to 1980, considerable federal government funds were provided to the states for increased agricultural extension services. The rationale for such public funding was not necessarily a failure of markets to provide extension services (at least, of an advisory nature) but rather that the system of markets would not bring about the socially-desired rate of growth that could potentially be achieved by increasing agricultural productivity. After 1980, general extension grants were discontinued, and increased federal funds were directed to research for rural industry. Until recent times there was lip-service to technology transfer under this new scheme. More recently technology transfer has been specifically recognized as part of this funded research process. Rather than being broadly-based, such extension is industry- and problem-focused.

The paradox of effective technology transfer, wherein more complex systems of farm management require one-on-one dealings between extension workers and farmers and such dealings convey private goods to farmers, is, in part, nicely solved in Victoria's dairy industry *Target 10* program. The Dairy Research and Development Corporation, with equal farm commodity tax and government grant contributions, has provided funding for government-employed extension officers to promote technologies to increase dairy production efficiency. A consequence of farmers contributing to the funding of the Target 10 program has been a significant level of farmer involvement in the planning and implementation of the program—something normally not characteristic of traditional government delivered extension programs in Australia.

In southeastern Australia a novel pilot private extension delivery project, the Farm Management 500 program, has been established. This program is financed by fees paid by member farmers plus commercial sponsorship by groups of input suppliers. In this program, which commenced in 1992, farmer groups are provided with non-advisory, educational extension services by a consortium of privately employed agricultural consultants with partial financial support from rural sector commercial sponsors. In 1994 the program comprised 450 farmer members, 15 privately employed consultants, and five commercial input or service supplier sponsors. Farmer membership fees contributed about half of the operating expenses. In 1994-95, farmer members paid about AUS $350 (1 $ U.S. = 1.15AUD at time of publication), as did the sponsors. It is a debatable question whether farm members would be willing to pay the full, unsponsored, cost. Potentially, such groups can operate efficiently on a large scale, with coordinated extension objectives. Particular entrepreneurial skills are required to initiate and maintain commercial sponsorship for such programs on an on-going basis and to market the program to farmer clients (Cary and Wilkinson 1994).

CONCEPTUAL ASPECTS OF PUBLIC VERSUS PRIVATE PROVISION OF EXTENSION SERVICES

The characteristic of privatized extension systems is a focus on commercial provision of services to farmers. When extension is delivered privately it represents a commercial decision; when extension is delivered publicly it is a political or bureaucratic decision. Both approaches have their advantages and disadvantages and, of course, their proponents and detractors.

In establishing the basis for public versus private extension we must first acknowledge that the nature of activity funded from public revenues is essentially a political decision. Often it reflects insufficient political will to change a decision made in earlier circumstances. Such decisions may be justified, or rationalized, as supporting the development of local industry, providing employment, or encouraging export-focused industry. In earlier periods in Australia, agriculture was often seen as fitting these criteria and accordingly received government support and infrastructure development (see Barr and Cary 1992). In the present era, leisure and service industries seem more likely to attract government support. Nowadays, governments in Australia are more likely to make grants to fund construction of motor racing circuits which attract international competitive events. In a similar vein, in the United States some governments have contributed to construction of major event football stadiums.

In determining whether to privatize it is important, in the first instance, to establish whether an extension program is designed to help commercial enterprises or is designed to help small scale farming and rural development. In the latter cases, privatization will be problematic and commercial provision is unlikely to occur without government subsidy (for example, public funding of private delivery).

While privatization is often contended to be a more efficient means of service delivery, reduction of government outlays is frequently an implicit, if not explicit, reason for its espousal. Thus, there are two themes in the broader privatization debate: a political economy consideration of the role and size of government, which focuses on whether or not there is a failure of private markets, and an expressed need to reduce government outlays.

There is an additional case where governments consciously sponsor extension to counter social disadvantage. Such circumstances are not common in Australia or New Zealand. Some examples in Australia are the public funding of rural counselors associated with regional economic hardship, and the government subsidization of commercial management planning advice as part of the Rural Adjustment Scheme. Having acknowledged the political reality and putting to one side questions of equity possibly associated with regional social disadvantage, we can consider a 'rational' approach to decisions regarding public or private funding of extension.

In mixed economies the prevailing economic justification for government involvement in an activity, such as agricultural extension, is because of market failure whereby the market mechanism alone cannot perform all economic functions for appropriate resource allocation. Market failure may arise because some goods or services are public goods (such as publicly-funded agricultural research knowledge) which can be consumed or used in a non-rival fashion by all members of society without any individual's use reducing the amount available for others. Because individuals cannot appropriate the benefit of providing such goods, they generally will not provide such goods in a society even though there may be significant gains for producers and consumers. Some extension activities are clearly concerned with public goods subject to market failure; other activities (such as individually-tailored advice) confer appropriable private benefits which could be adequately supplied by private markets.

Private goods are also subject to market failure when the operation of private markets does not provide certain services at a socially optimal level, where external costs or benefits are accrued by others rather than the provider of the goods, or when current generations place insufficient value on preservation of resources for future generations. Publicly funded conservation extension is often directed to overcoming this last type of market failure.

The argument for extension privatization is concerned with more efficient delivery of services (because of the greater efficiency of private markets for services) and lowered government expenditures. Efficient markets quickly resolve questions of supply and demand by swapping information between users and suppliers of extension services about what services are required and what might be supplied, as well as feedback about service provision. The argument for public extension is that much agricultural information is a public good (or a mix of public and private good that is difficult to separate); and that there is a need for better management of common property natural resources, as well as to provide for farmers who lack access to educational or management services.

In some cases, agricultural extension services convey nearly exclusively private goods (e.g. specifically-tailored farm management advice); other situations clearly are characterized as public goods. There is a lot of fuzzy ground in the middle where it is not particularly clear whether an extension activity is conferring a public good or private good (Watson et al. 1992). In such situations the 'ideal' approach has to be determined on the merits of the case. At least in theory, the public funding and private delivery of certain extension services (such as by outsourcing) may allow increased efficiency of delivery for situations which have public good characteristics. In practice, the process of making grants for such situations, if not frequently scrutinized and monitored for quality suffer the inefficiencies inherent in 'grants economies'.[2] Government funding by budget grant (for either public or private delivery of a 'service') often involves a budget allocation determined at an upper level in a bureaucratic or political hierarchy, with the administration and monitoring of service delivery at a lower administrative level. The administrative level of delivery often has a vested interest in maintaining the grant regardless of its usefulness or the efficacy of its delivery (Boulding 1981).

Allied to the 'efficiency' argument is the potential of more direct accountability between extension provider and client. Unlike the U.S. county agent system, in Australia the delivery of agricultural extension through state Departments of Agriculture has resulted in control via centralized government bureaucracies. Extension work was planned in conjunction with head office personnel and only a small participation by local farm people (Williams 1968). This has generally led to a low level of direct accountability to 'clients' and a somewhat paternalistic approach. In the past the monolithic approach of large 'monopoly' government agencies is likely to have inhibited innovation in extension service delivery.

TECHNOLOGY TRANSFER

Typically in industrial firms research and development activities have institutional links with the corporation's marketing activities which provide the 'technology transfer'. In such situations the question of 'who should disseminate?' rarely arises. In the cases of agricultural research where the research is a public good, this can be a moot question. It ought at least be clear that public agency extension does not need to focus on disseminating technologies which are the domain of 'private' industry. Some Australian and New Zealand examples of the legitimate use of public and partially private or industry funding for either public or private delivery of tailored technology transfer have been discussed above.

Not all extension expenditure can be assessed against the benefits from technology transfer: the benefits of extension concerned with human development are difficult to quantify in the short term. In developing regions, or socially disadvantaged localities, extension often substitutes for broader systems of education and for other forms of regional and community infrastructure.

As farm businesses have become larger, technology transfer now occurs against a background of an increasing need for more complex farm management advice as well as technical advice. It has been a difficult challenge, and potentially problematic, for public extension agencies to adequately frame traditional *technology* transfer within the demands for advice which take account of individual management contexts.

PUBLIC GOODS AND RESOURCE CONSERVATION

Within the last two decades there has been a surge in public concern over the conservation and restoration of environmental resources (Dunlap 1992; Barr and Brown 1994). The rise of modern environmentalism has coincided with increasing public support for privatizing the delivery of government services with private good characteristics. The conservation of environmental resources typically embraces public goods characterized by external costs and benefits accruing beyond a particular site, or costs and benefits which may be incurred by a future rather than the present generation. Public discussion of rural environment conservation and of privatization is often, and sometimes purposely, confused. Thus it is important to understand the relevant conceptual issues discussed earlier.

In countries which do not produce artificial food surpluses it is likely that the short or medium term economic returns to government investment in conservation and restoration of rural environmental resources will be less than an equivalent investment in technology transfer related to production services. Hence public good investments in resource conservation will depend on public or government commitment to the 'public good'. It is also axiomatic that more affluent countries will find it less difficult to invest in such public goods than will less affluent counties.

It is worth considering how environment and ecosystems are likely to fare under the various faces of privatization. Governments committed to privatization are often principally committed to reducing government outlays; in such cases government support for resource conservation and restoration is likely to be curtailed. More importantly, because there will be no appropriate private market to deliver such services, they will not be delivered by private individuals. More far-sighted governments will understand the public good nature of such activity and may well reduce production services that can be supplied privately in order to maintain or augment support for public good conservation services.

Traditionally, public good conservation extension services have been delivered by public agencies. Victoria has recently attempted to capture potential efficiencies of private delivery of conservation services by proposing the 'outsourcing' of environmental services, whereby private agencies tender to deliver public good conservation programs funded by government (see Table 1). The Department of Conservation and Land Management is currently developing a pilot program of contracting out the delivery of its extension services in two regions in Victoria. In New Zealand, while production orientated advisory services have been privatized, the delivery of natural resource conservation programs has been transferred from central government to regional authority ownership (Regional Councils); and funding for these programs is contributed by levies on regional landholders, thus attempting to better match regional, if not local, external costs and benefits. Where a recognizable group can be identified as the principal beneficiaries from government provision of public goods, specific taxes or levies provide an equitable basis for funding such activities.

In Australia there has been a rapid growth of landcare groups. Landcare groups are voluntary community land conservation groups. Landholders in a rural locality work together to tackle local land protection problems such as salinity, soil erosion, rabbit infestations, or tree decline. The groups identify local land degradation problems, create awareness about them, and undertake works (such as tree planting) to overcome them. The use of local groups overcomes some of the local externalities associated with equitably allocating costs and benefits between neighboring landholders, whereby one property holder may incur disproportionate costs of land restoration and another may receive disproportionate benefits. Involvement of land users in solving local problems has enhanced local understanding of the problems of land degradation and the development of locally appropriate land management systems. In announcing the 'Decade of Landcare' the Australian federal government hoped there would be 1,200 Landcare groups by the year 2000, a target which was exceeded by 1993. In 1995 there were approximately 2,700 Landcare groups. In Victoria, approximately one third of farm properties have a member of a Landcare group (Mues et al. 1994).

The enthusiastic embrace of Australian landholders for the landcare movement is a phenomenon that has not occurred in New Zealand; nor does it appear to have an equivalent in other western countries. The explanation for this Australian phenomenon is likely to be complex. Australian governments have been wary of implementing legislative controls on rural land use practices for fear of damaging export competitiveness. The predominantly urban electorate has generally had little interest in providing significant subsidies to encourage changed practices on farms. Consequently, most Australian natural resource conservation programs rely predominantly on voluntary compliance or the informed self-interest of land users. In 1983 spectacular dust storms briefly engulfed cities in south eastern Australia, a consequence of an extended drought in the early 1980s. Urban people, in an era of heightened environmental awareness, were thus sensitized to problems of rural land degradation. This community concern has been maintained and accelerated into the following decade, and has been accompanied by large government contributions to Landcare activities and delivery infrastructure. Urban community support, together with an unusual coalition of interest between the national farmers' organization and the Australian Conservation Foundation, has ensured the growth of the program.

The role of government agents in Landcare has been primarily as technical consultants and as group facilitators—in both cases providing a public good extension service. Government extension officers also administer voluntary programs to encourage biodiversity (such as Land for Wildlife in Victoria and Tasmania, and conservation on private land in Western Australia). In South Australia, extension officers often supervise voluntary management agreements involving government-provided incentives to encourage appropriate environmental behavior (Clairs and Young 1995).

Public goods associated with resource conservation and restoration are usually more judiciously assessed when assessment is made from the perspective of distance. For example, local people rarely see the wisdom of declaring local land as a national park. An interesting question which arises with respect to the New Zealand, in contrast to the Australian privatization experience, is whether regional funding and delivery of conservation and restoration activity results in more acuity in recognizing and tackling regional land conservation problems or a heightened reticence to face the costs of tackling such problems.

CONCLUSION

There are two major themes that arise from any comparative consideration of the changes in extension delivery that have occurred in Australia and New Zealand. First, the advisory focus of New Zealand extension was more amenable to the change to commercial consulting. Second, in New Zealand the change from government agency to fully privatized business extended over ten years, during which time government resolve to implement the change remained firm. The series of adaptations which occurred over this period embraced both institutional learning and important adaptations by personnel learning to operate within a commercial environment

There are a number of conditions which can be considered necessary or conducive for a transition to a fee-based, or privatized, system of technology transfer.

The service in question should have clear private good characteristics, such as many advisory services provided directly to individual farmers. Modern industrialized agricultural sectors using more sophisticated purchased inputs require sophisticated and more specialized management systems. Public systems of technology transfer will be unlikely to adequately or equitably deliver services in these circumstances. Agricultural sectors in industrialized economies with well-developed infrastructure and better educated farmers, geographically concentrated and integrated with both public and private institutional knowledge systems, are well disposed for adoption of fee-based systems. Where such conditions are not present, commercialization or privatization is more problematic. For a transition to a fee paying service, ideally, a nucleus of private deliverers of advisory services should already exist to provide models for commercial delivery. Public good conservation services are likely to remain within the sphere of government, although, as discussed above, governments may seek to use private systems of delivery while providing public funding.

Those extension services that have adopted a commercialization or privatization strategy most vigorously, traditionally have employed an advisory approach to extension delivery. The advice given is more likely to be a private good, and more easily and naturally lends itself to privatization. As well, the extension advisers are more likely to be able to adapt to providing services commercially. However, some staff will not make such a transition easily; new commercial skills will be required by newly commercialized advisers, and the dynamics of any change will have to be planned carefully.

A particular difficulty for public organizations seeking to privatize is evident from the New Zealand experience. Most private professional businesses tend to commence from small beginnings as a small organization with one or several professionals who trust and respect each other. New colleagues or partners join the business as demonstration of competence and demand for work dictates. In New Zealand some district or regional groups of extension advisers had these characteristics, but the agency was offered for sale 'as a whole.' Because in any such organization there will be personnel who are unproductive or unsuited to commercial imperatives, and because such information will not be well distributed through the organization, it is usually difficult for staff in a large organization to determine a reasonable buy-out price. This may be overcome if there has been an extended period of commercial delivery in which the commercial competence of personnel has been established and identified and, by implication, non-performers have withdrawn from the organization.[3] Governments normally want to 'wash their hands' of such personnel problems by divesting an organization quickly.

Fee setting is an important consideration for an agency commencing on a path to privatization. There is a danger that a government agency charging for its services will unknowingly, or perhaps knowingly, set fees for services which represent subsidized service or predatory pricing. Le Gouis (1991) has noted that government 'commercial' fees should be set at the market rate so as to not compete unfairly with existing private consultants or to prevent establishment of new consultants. The potential for free publicly provided services to 'crowd out' privately provided services can be observed in the recent experience in the north west of Victoria: declining government extension services have been replaced by an increase in privately delivered services.

The Australian and New Zealand experience can be used to identify the important issues of public versus private technology transfer and provide some lessons for institutional change in extension delivery and technology transfer elsewhere. There is no particular reason why the balance between public and private delivery, as it exists at one time, should be the balance that exists at another time. Dynamic forces which militate against constancy are at work in agriculture and the wider economy.

NOTES

[1] Agriculture New Zealand's most valuable asset was probably the goodwill associated with its professional staff. Staff, understandably, did not wish to pay highly for their own 'assets', which individual consultants could retain by commencing business independently. The Agriculture New Zealand staff engaged an international financial adviser to recommend a bid price. The valuation was less than Treasury placed on the business. The price at which the business subsequently was sold to Wrightson Ltd. has not been made public.

[2] The provision of public goods is generally provided by grants rather than by direct commercial exchange (hence, a grants economy in contrast to an exchange economy).

[3] This situation would be more like a 'management buy-out' in a commercial business. In New Zealand between 1986 and 1992 consultants who could not deliver the fee level required by the business or who chose not to meet the commercial imperatives generally left the business. Such a period allows a transition particularly for technically skilled people between 40 and 50 years of age who may have difficulty adjusting to the commercial requirements. In the New Zealand transition, about half the staff in this group became effective consultants and about half resigned to find new careers.

REFERENCES

Barr, N.F., and M. Brown. 1994. Landcare: Sustaining public opinion. Paper presented to the Australian Landcare Conference 'Landcare in the Balance', Hobart, Tasmania, September.

Barr, N.F., and J.W. Cary. 1992. Greening a Brown Land: The Australian Search for Sustainable Land Use. Melbourne: Macmillan.

Boulding, K.E. 1981. A Preface to Grants Economics. New York: Praeger.

Cary, J.W. 1993. Changing foundations for government support of Agricultural Extension in economically developed countries. Sociologia Ruralis 33: 336-347.

Cary, J.W., and R.L. Wilkinson. 1992. The provision of government extension services to the Victorian farming community. Parkville, Victoria: University of Melbourne, School of Agriculture and Forestry.

Cary, J.W., and R.L. Wilkinson. 1994. An Evaluation of Farm Management 500: A Programme Promoting Better Business Management. Parkville, Victoria: University of Melbourne, School of Agriculture and Forestry.

Cary, J.W., and R.L. Wilkinson. 1995. The use of publicly-funded Extension Services in Australia. Journal of Extension 33(6).

Clairs, T., and M. Young. 1994. Approaches to the Use of Incentives to Conserve Biodiversity. Canberra: CSIRO.

Dunlap, R.E. 1992. Trends in public opinion towards environmental issues: 1965-1990. In American Environmentalism: The U.S. Environmental Movement, R.E. Dunlap., and A.G. Mertig (eds.), 1970-1990. New York: Taylor Francis.

Harrison Market Research. 1991. Survey of farm managers: study overview and commentary about commercialization. A report to the Department of Agriculture, South Australia.

Hercus, J.M. 1991. The commercialization of government agricultural extension services in New Zealand. In Agricultural Extension: Worldwide Institutional Evolution and Forces for Change, W.M. Rivera and D.J. Gustafson (eds.). Amsterdam: Elsevier.

Kloppenburg, J.R. 1988. First the Seed: The Political Economy of Plant Biotechnology, 1492-2000. Cambridge: Cambridge University Press.

Le Gouis, M. 1991. Alternative financing of agricultural extension: Recent trends and implications for the future. In Agricultural Extension: Worldwide Institutional Evolution and Forces for Change, W.M. Rivera and D.J. Gustafson (eds.). Amsterdam: Elsevier.

Mues, C., H. Roper, and J. Ockerby. 1994. Survey of Landcare and land management practices, 1992-93. Canberra: Australian Bureau of Agricultural and Resource Economics.

Neilson, D. 1992. Institutional reform and privatization in research and extension. Unpublished paper, presented at Latin American Agricultural Retreat. Harpers Ferry, WV, September 8-10. Washington, DC: The World Bank.

Paarlberg, D. 1978. Agriculture loses its uniqueness. American Journal of Agricultural Economics 59: 769-772

Ritchie, I. Undated. From the public to the private sector: The Agriculture New Zealand story. Unpublished paper.

Rivera, W.M., and J.W. Cary. 1995. Privatizing agricultural extension. In Improving Agricultural Extension: A Reference Manual. Rome: Food and Agriculture Organization, B.E. Swanson (ed.). In press.

Walker, A.B. 1993. Recent New Zealand experience in agricultural extension. Australia-Pacific Extension Conference Proceedings, Vol. 1, Brisbane: Department of Primary Industries, pp. 126-129.

Watson, A.S., R. Hely, M. O'Keeffe, J.W Cary, and N. Clark. 1992. Review of Field-based Services in the Victorian Department of Food and Agriculture. Melbourne: Agmedia.

Williams, D.B. 1968. Agricultural Extension: Farm Extension Services in Australia, Britain and the United States of America. Carlton: Melbourne University Press.

11

Landcare: A Recent Australian Extension Phenomenon

Brian Scarsbrick

Soil degradation and water pollution are the most important environmental issues facing Australia. More than one half of Australia's farmland requires treatment for land degradation and it is costing the economy well over AUD $1 billion (1 $ U.S. = 1.15 AUD at time of publication) a year in lost production alone (Prime Minister's Science Council Booklet 1995).

The total cost of land degradation, including treatment of degraded land, nutrient loss, research, and costs related to silting and pollution of rivers, lakes, dams and harbors, would be much higher—perhaps AUD $2 billion a year. Directly or indirectly, it is a cost that our whole community bears.

Half the major forests and about 35 percent of Australia's woodlands have been cleared or severely modified. Much of our crop land has been pulverized into submission resulting in various combinations of soil erosion, salinity, acidification, soil structure decline, and water logging now affecting a significant proportion of the land used for agriculture (Campbell 1993).

Most forms of land degradation are natural processes accelerated by human activity (Roberts 1991). For example, it is estimated that soil erosion in some of the important agricultural regions of Australia is up to fifty times the natural rate (Chamala and Mortiss 1990).

Put simply, for every loaf of bread being produced, an average of seven kilos of soil was being lost from our crop lands under conventional cultivation. For every bottle of wine, one kilo of soil was lost to our rivers and oceans.

ISBN 1-57444-104-3/98/$0.00/$.50

The quality of the water in our streams and rivers is a real litmus test on how well the land within the water catchment is being managed. Soil, water, vegetation, and animal/human activity are inextricably linked and their interaction dictates whether the management of the land is ecologically sustainable. The dramatic increase in the incidence of blue green algae blooms in major catchments such as the Murray-Darling and Hawkesbury-Nepean is stark evidence of water quality problems facing Australia. Two years ago a thick ribbon of blue green algae wound 1,000 kilometers from Mungindi to Wilcannia—it was the biggest algal bloom ever recorded on any river in the world. It was this event that focused the nation's attention on the quality of our water. Over Easter 1992, holiday makers were sent home from a popular dam near Brisbane—the reason, fear of blue-green algae toxin. The Chaffey Dam in northwestern New South Wales has for the past four years been plagued with blue-green algae blooms, restricting its potential to supply Tamworth with quality water. With blue-green algae toxin 10 times more poisonous than strychnine at its worst, and recent research indicating that sublethal doses of the toxin cause cancer in experimental rodents, it is little wonder that water quality is seen by the Australian public as the highest priority issue on the environmental agenda (Australian National Opinion Poll Survey 1991).

In the words of Bob Beale, editor of the Sydney Morning Herald and author of *The Vanishing Continent*— "no other environmental issue so plainly affects every single Australian. It ravages the economy, it fouls rivers, harbors and drinking waters, it threatens livelihoods, destabilizes whole communities and casts a long shadow over future generations."

Land degradation is an insidious disease and it is threatening to kill Australia.

LANDCARE–A PARTNERSHIP

The strength and rapid growth of landcare across Australia is based on partnerships. In 1989 the National Farmers' Federation and the Australian Conservation Foundation (normally an unlikely partnership) got together and lobbied the then Prime Minister, Bob Hawke, to support a community group movement started in the state of Victoria called Landcare to "turn the tide" on soil and water degradation. As a result, the Decade of Landcare was announced and the government pledged AUD $32 million (now AUD $107 million) a year for ten years.

The federal government, in partnership with state governments and the community, encouraged the establishment of landcare groups across Australia to voluntarily change their land management practices and share information on sustainable land use.

Landcare Australia has created partnerships between Landcare and the corporate sector. Sponsorships for awareness programs and landcare group projects provide valuable exposure for the sponsors and practical help to the groups involved.

GRASS ROOTS ACTION

The landcare movement is growing rapidly throughout Australia as a grass roots movement for achieving sustainable land management. There are now almost 3,000 landcare support groups who are seeking to change land use practices. Close to 35 percent of our farmers are actual members of the landcare group network, and many others are individually seeking to adopt sustainable practices to help reduce land and water degradation. A typical rural landcare group is made up of a number of land managers (farmers/grazers) in a localized area meeting together to solve common problems. The following are two examples of landcare group activities:

- *In New South Wales the Boorowa Community Landcare Group is working to reverse land degradation and water quality problems.* 'The sewer of the Lachlan River' was the sobering finding (from the combined research of the Yass Soil Conservation Service and Forbes Department of Water Resources) that confronted the Boorowa Community Landcare Group in 1989. The Boorowa River had been identified as a major contributor of salt into the Lachlan River system. It was the result of the state of the surrounding landscape which was suffering from rapid deforestation, increasing soil acidity, declining soil structure, the emergence of dryland salinity and the appearance of features associated with rising water tables. The stark reality was a very real threat to sustainable and viable agriculture in the Boorowa River catchment area. The group has achieved early success with tangible production on previously barren discharge areas. They are optimistic that the days of the Boorowa River being the 'sewer of the Lachlan' are numbered.

- *In Victoria the Bass Valley Landcare Group is working in the Gippsland area.* The Bass Range in South Gippsland, Victoria, was almost totally cleared for dairying in the 1890s. As a result the area now shows several types of land degradation: tunnel erosion, land slips, rabbit and weed infestations, and stream pollution from dairy effluents. The Bass Valley Landcare Group is probably the largest in South Gippsland, with over 180 farming families involved. To tackle the issues in the area and improve water quality, the

> group has pioneered close working relations with regional water boards and encouraged farmers to better manage effluent, create artificial wetlands, and revegetate streamsides. Roadside weed control programs in conjunction with the State Roads Authority and local councils have been outstanding and Bass Valley School's junior landcare group is enthusiastic and productive.

The current rate of growth of landcare does not appear to be slowing down as the concept of a "grass roots driven, middle of the road" community action movement gains acceptance with the mainstream farming community. The urban population is also starting to recognize their contribution to the problems and what they can do about it. Urban landcare groups are being formed at an increasing rate to address issues such as sand dune stabilization (Coastcare), bush regeneration (Bush Care) and water quality monitoring (Waterwatch).

The National Landcare Program developed by the federal government in cooperation with state governments and the community is unique. The landcare groups actually identify their local problems and can apply for Commonwealth funding (up to AUD $20,000 per year) to collectively work to achieve solutions to the land degradation problems. It is this "bottom up" approach that is driving the movement and a major reason for its success. Over AUD $100 million per year is now being contributed by the federal government for the Decade of Landcare. Each state is also making substantial contributions to landcare, and corporate sponsorship for awareness campaigns and group projects is also being provided by Landcare Australia.

In very few places in the world is there a network of community groups spearheaded by farmers that is actually delivering results "on the ground" to improve soil and water quality. Landcare works because it fosters community spirit involving farmers, schools, companies, government and even the urban population.

LANDCARE AUSTRALIA LIMITED

As part of the Decade of Landcare initiative the federal government also established Landcare Australia Limited, a non profit public company, to undertake two main tasks: to develop a landcare ethic in the whole community through education and promotion; and to work with the corporate sector to raise funding for landcare projects. Landcare Australia has on its board of management representatives from the corporate sector, the Australian Conservation Foundation, the National Farmers' Federation, State Departments responsible for natural resource management, the Commonwealth Scientific and Industrial Research Organization, and the media. This diverse membership allows the company to work in partnership with the corporate sector to achieve its objectives.

Landcare Australia does not distribute government funding but assists groups with corporate sponsorships, information flow, and awareness across the nation. For example, a sponsored national magazine "The BP Landcare Challenge" goes to all groups free of charge promoting what landcare groups are achieving in each state.

RURAL COMMUNITY ACTION FOR CHANGE

Property management planning

A significant tool to achieve sustainable agriculture is the adoption by landholders of property management planning. Property management planning involves property resource and risk assessment, financial planning, smarter marketing and even setting social goals for the farm family. Landcare seeks to provide training to enable landholders to develop their own property management plans. These plans may include strategies such as improving subdivision so that land use is based on land capability rather than arbitrary boundaries; increasing on-farm vegetation cover with trees and improved perennial pastures to reduce salinity, soil loss and acidification; treatment for soil erosion; and the adoption of no till practices, particularly in areas that have more fragile soils.

Landcare and water catchment management

The landcare group is the basic unit of community action that cumulatively can solve national land and water problems. The landcare group approach offers the opportunity to tackle huge and often daunting national problems in 'bite size chunks' and find local solutions which are driven by grass roots concern for the future of our natural resources.

In the smaller catchments, a landcare group would ideally contain all properties within the catchment. Each landholder would be encouraged to develop a whole farm property plan to achieve sustainable productivity. These property plans would be integrated to develop a total water catchment plan. The huge potential for improvement in soil and water quality is in a catchment covered by a landcare group whose membership consists of the actual managers of the land working together to achieve a common purpose.

Catchment management committees are being formed in most states. In larger catchments, a number of landcare groups would join together and a water catchment plan could be developed using a catchment management committee to coordinate the process. For example, in New South Wales, the whole state has been divided up on a water catchment basis and Total Catchment Management Committees have been formed to coordinate activities.

Most states are organizing the management of natural resources on a catchment or regional basis. In this way local landcare groups that are represented on the catchment committees can be the driving force that implements a national landcare strategy ensuring that local knowledge and input is maximized.

The success of landcare as a change agent

The landcare group movement provides a model for local development of management practices which improve soil health and water quality. The adoption of no till practices and conservation farming has been rapid in some groups, and the opinion leaders in the group have often acted as advisers to other members. A survey was undertaken by the Australian Bureau of Agricultural and Resource Economics (ABARE 1994) comparing farmers who were members of a landcare group with those who were not. ABARE found that landcare group members have larger farms, higher levels of farm cash income, and, in spite of higher debt load, had equal or higher returns to capital than non-members. Landcare members score better than non-members in the adoption of more sustainable farming practices such as:

- the separation of different land classes for different uses (51 percent of members do this compared with 33 percent of non members)
- minimum tillage (59 percent vs. 40 percent)
- direct drilling (37 percent vs. 21 percent)
- excluding stock from degraded areas (47 percent vs. 25 percent)
- managing crop rotations to minimize degradation (62 percent vs. 42 percent)
- stubble retention or mulching to minimize degradation (46 percent vs. 37 percent)
- placing watering points to minimize degradation (44 percent vs. 29 percent)
- regular soil testing (40 percent vs. 28 percent)

So it can be seen that not only is the movement gaining in members but also the promotion of the ethic is having an effect, and landcare group members are moving to more sustainable agricultural techniques to a much greater extent that non-landcare group members.

More than one million tons of nutrients are exported from Australia each year in agricultural produce. Maintaining the correct nutrient balance by soil testing and applying the correct rates and types of fertilizer, as well as crop rotation and stubble retention, ensures that nutrient depletion from the farming system is minimized. Over grazing is a major cause of land degradation and maintaining the correct vegetative cover to ensure that production does not fall over time (because of land degradation) is of paramount importance. There are a number of examples where landcare groups have developed whole catchment plans to address the problem of dryland salinity. Areas of high recharge in the catchment have been revegetated and streambank erosion reduced by creating buffer strips of vegetation along river banks to minimize nutrient and sediment deposit into rivers.

The basis for control of many salinity problems involves development of appropriate Best Management Practices (BMP). Many vegetation and pasture management practices can lead to improved productivity while reducing the groundwater problems that are common to both dryland and irrigation agriculture. Landcare groups using aerial electromagnetic survey techniques to identify saline "hot spots" and undertaking strategic revegetation is an example of BMPs being adopted in dryland catchments.

In the case of irrigation, these management practices may be supplemented by activities aimed at reducing water usage. Improved irrigation technology, laser grading, and water recycling all contribute to water conservation. Improved management practices will ensure a reduction in salinity of local and regional water and ease the threat of widespread salinization (National Water Quality Strategy 1992).

GOVERNMENT EXTENSION SERVICES

Traditionally agricultural extension services have been provided by state governments and to a lesser extent private enterprise. Universities have not normally provided extension services and have relied on the state agencies and private consultants to extend their research results to the farming community.

Increasingly, state government departments are cutting back on their extension services and are making the distinction between advice that will yield substantial public benefit compared with advice that would provide a private benefit to individual landholders.

In the main, state departments of agriculture and conservation (Victoria, South Australia, Western Australia, and Tasmania) are not providing advice to individuals (property visits) and are restricting their activity to group extension and developing innovative information transfer systems.

Budget cuts have also reduced the ability of the government agency driven extension to adequately service their clients on an individual (face to face) basis. Face-to-face advice to individual farmers which is likely to lead to a private benefit is being left to private enterprise

The landcare group network is the mechanism through which agricultural extension, natural resource management, information and technology transfer is being delivered both by the public and private sector. For instance, the Western Australian Department of Agriculture has developed an automated fax back system to provide a 24-hour information service on prepared Agfax documents on current issues.

PRIVATE SECTOR DELIVERY OF EXTENSION

The private sector has been providing extension services to the agricultural community for many years. Private consultants are commonly providing advice on issues ranging from whole farm management to specifically tailored advice such as Integrated Pest Management. Corporate farms are developing their own internal expertise. Agribusiness (i.e., input vendors) is providing an advisory service that is often tailored to their marketing objectives.

GOVERNMENT CONTRACT EXTENSION

The Victorian Government is moving to contract out to private enterprise the delivery of natural resource management advice on a water catchment basis. Two trial catchments—Maribyrnong/Werribee and the Campaspe—have been selected by the Victorian government to tender for private enterprise to deliver the landcare facilitation service in these areas.

The Department of Conservation and Natural Resources in Victoria is currently drawing up the specifications and going through the tender process as an alternative to government providing a total service. The current role that state governments play in facilitating the development of landcare groups (e.g., employing facilitators) will be contracted out on an experimental basis. This may be a forerunner for the delivery of extension services in other catchments.

GROUPS MOVE TOWARD PRIVATIZATION

A number of landcare groups have developed skills and expertise in natural resource management and are now seeking to generate income by marketing their expertise. Landcare groups such as the Jerramungup Land Conservation District Committee in Western Australia, have developed GIS technology which has been used for property management planning for their group members. The group is now in a position to diversify and undertake tasks on a consultant basis for the local shire (road maintenance programs), Department of Agriculture (soil survey mapping), and other landcare groups.

Kondinin Group, a farmer group formed 40 years ago in Western Australia, has developed rapidly in recent years and now has a membership of 8,000 farmers and others in the agricultural business and it now markets agricultural information in an innovative way on a user pays system. Apart from providing a regular magazine and a series of books on current topics, the group has now developed "Farmline." Farmline is a national phone or fax information service that guarantees answers to specific questions from farmers or agricultural service industry members within 24 hours. This "not for profit" farmer group provides information on crops, livestock, pasture, machinery, insurance at a reasonable cost and is becoming popular.

Farm Advance is a farmer group which employs its own coordinators to ensure that the group members are kept up to date with the latest agricultural and natural resource management information. Apart from a regular newsletter, farm field days, and workshops are organized with the best expertise brought in to address the group.

RESEARCH AND LANDCARE

Landcare groups are undertaking farm research, usually in association with a research institute, and are sharing this information with landcare group members. Researchers are turning to landcare groups to provide a nucleus of farmers to develop "on farm" research and disseminate their findings through group involvement. Landcare groups are able to try out "best bet" sustainable agriculture practices before they are promoted to the wider farming community.

AGRIBUSINESS AND LANDCARE

Companies which regard the rural market as important clients are finding that the landcare group network is an ideal vehicle for their advertising and relationship marketing activities. Companies such as BP Oil, Combined Rural Traders (CRT) and the fertilizer industry, are strong sponsors of landcare, and those companies are developing market loyalty within the landcare network.

Just as the landcare group network is achieving success in encouraging the adoption of new more sustainable agricultural practices (ABARE 1994), agribusiness is finding that the uptake of new technology and products is also facilitated by the group process. Opportunities also exist for agribusiness to take over the delivery of landcare services to the groups as state government extension services are being withdrawn. Property management planning could be delivered by agribusiness in the same way as the soil testing service has been privatized.

INFORMATION TRANSFER AND COMMUNICATION

The channels of communication between landcare groups are somewhat tenuous and at this stage are not regarded as efficient. A new project is being developed at the national level linking into rural information "nodes" to improve information flow. It is intended that this communication system be seen to be owned by the landcare movement to ensure regular participation. A national magazine, the *Landcare Challenge*, sponsored by BP Australia, goes directly to most members of the landcare movement which assists with the information flow between landcare groups. Additionally, a three-year trial has just been completed (LandcareNET) which linked a number of landcare groups on a computer bulletin board system to share information. While a number of problems were identified such as a degree of user unfriendliness, user reluctance to participate, difficulties with phone lines in rural areas, etc., it is clear that the information super highway (Internet) will eventually be a major provider of information to the rural community.

THE RURAL ADJUSTMENT SCHEME

The Rural Adjustment Scheme was established to assist marginal landholders to adjust out of agriculture and the more efficient producers to increase the size of their holdings and gain economies of scale. The scheme provides for an interest rate subsidy to the landholders increasing the size of their holdings and a "safety net" (up to AUD $45,000) and training for those adjusting out of agriculture. Closer interaction between the landcare movement and the RAS Scheme would enhance its delivery.

THE RURAL/URBAN INTERACTION

Awareness campaigns at national and state levels such as the Uncle Tobys Company Landcare advertising campaign, National Landcare Month, and the Landcare Schools Video Competition, are penetrating through to the wider community as the attached Roy Morgan Research indicates (Figure 1). The research suggests that between July 1991 and September 1995 the awareness of landcare more than tripled from 22 percent to 69 percent on a national basis.

These results demonstrate that the national and state awareness campaigns are helping to foster a landcare ethic in the whole community. In capital cities, Landcare Australia's prime target, the awareness of landcare significantly increased from 10.6 percent to 63.2 percent.

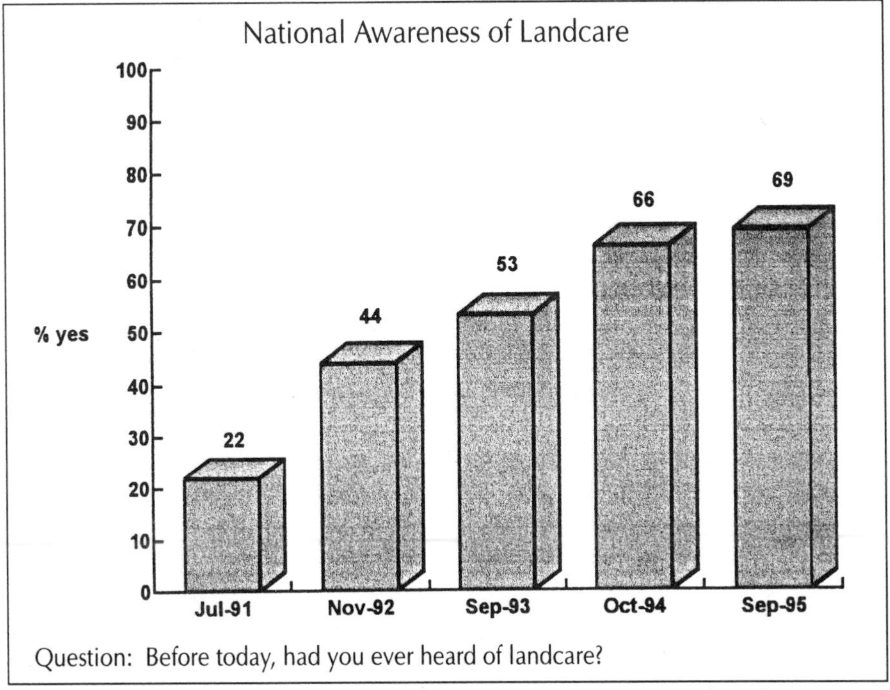

Figure 1: Results of poll taken by the Roy Morgan Research Centre Pty Ltd in July 1991, November 1992, September 1993, October 1994 and September 1995.

Rural and urban land and water management are closely linked, as the blue-green algae problem clearly demonstrates. Blue-green algal blooms are caused by a number of factors, but the increase in phosphate levels in water and sediments has been a major contributor to the problem. The New South Wales Blue-Green Algae Task Force (1992) reported that 50 percent of the phosphate entering streams results from run off from rural lands with the other 50 percent coming from urban sources. Sewage treatment plants contribute an estimated 50 percent of the phosphates going into rivers. For example, there are 171 plants in the Murray Darling Basin, with only three removing phosphate and a number using land disposal. The cities and towns positioned along inland rivers are making significant contributions to the problem of water eutrophication.

Water quality is the issue which is most likely to gain support and action from both rural and urban communities.

CORPORATE COMMUNITY ACTION FOR LANDCARE

A significant development in Australia has been the involvement of companies providing sponsorship backing for landcare. Major companies such as BP Australia, BHP, Telstra, Ansett Airlines, and Ford Australia have seen the advantages of supporting this community based movement. With public concern mounting over environmental issues, company balance sheets are reflecting consumer loyalty to those firms promoting a cleaner environment. As Neil Shoebridge stated, "A concern for the environment is one of the criteria for being an acceptable marketer in the 1990s" (BRW May 1992).

Landcare provides companies with the opportunity to develop their relationship marketing strategies throughout Australia. Landcare is seen as the "middle of the road" environmental movement that enhances the company's image with the community. A survey on the awareness of Landcare Australia's logo indicates that 62 percent of the population reported that they had seen the logo before. In addition, when asked if they thought landcare was a worthwhile cause, 95.6 percent believed it was.

Landcare Australia can help companies achieve a positive environmental profile. Landcare Australia tailors projects to meet corporate marketing objectives and ensures widespread promotion of the company both through authorized use of the Landcare Australia logo and the volume of media generated. As a result, many leading Australian companies have accepted the landcare challenge to support the landcare movement. The Uncle Tobys Company (which manufactures breakfast cereal and snack foods) sponsor a series of 'Let's Landcare Australia'—thirty second television commercials, giving pointers on how to care for the environment.

The following are some comments from companies that are providing substantial sponsorship to Landcare Australia. "Uncle Tobys and Landcare Australia are working together to promote landcare issues, thus enhancing the quality of out environment," (Steve Dillon, Marketing and Sales Director).

Chuck Bowman, Managing Director of BP Australia (a major international petroleum company) launched a raft of proposals at the 1991 National Landcare Awards, including sponsoring a quarterly magazine and video to circulate information to landcare groups, plus financial support through BP distributors for local projects. "We used the BP Landcare Challenge as a marketing tool to get into an area we hadn't been in for five years. We've gone from virtually nothing to a 65 percent share and market leader (petrol). Before we couldn't get their business and now they're calling me. It's the best thing BP's ever done." (Scarsbrick 1992). Telstra Australia (a company that provides national telecommunication services) is sponsoring the "Local Government and the Education Awards" and several other projects. As Michael MaKieve, Executive General Manager, said "Landcare actually provides one of those rare opportunities for a win-win situation."

Ford Australia is sponsoring the Media Award. Ian Vaughan, Vice President, External Affairs, made this comment: "Ford Australia supports the activities of Landcare Australia in which government and non-government organizations work together to preserve our productive landscape. Landcare provides a balance between the needs of primary production and conservation in a way which is achieving results for the well-being of this country's environment." (Scarsbrick 1992).

An analysis of why companies sponsor landcare reveals that they regard landcare as a cause that is vital to the future of Australia, but just as importantly "because it's good business." (Loton 1995). The National Landcare Australia Awards, held biannually, are the promotional flagship of Landcare Australia. The awards acknowledge and reward landcare groups, companies, individuals, and the community for outstanding environmental achievements. Hundreds of entries are received in ten categories which recognize achievements in business, agriculture, education, individual efforts, and the media. The Landcare Awards ceremony is staged in the Great Hall at Parliament House Canberra and attended by dignitaries including the governor-general and Prime Minister. It is a prestigious event and receives national media coverage.

CONCLUSION

Landcare provides an innovative approach to combining upper level resources (government and private) with local knowledge and incentive. However, in addition to the ABARE survey data more research is necessary to properly evaluate and monitor the progress of landcare in achieving resource conservation and its contribution to the long term sustainability of Australia's ecological, economic, and social fabric. Reliable benchmark data needs to be established on the state of our soil, rivers, vegetation, and biodiversity. The progress of soil salinization and rising subsurface groundwater tables needs to be monitored nationally. Soil erosion, acidification, and soil structure decline requires national benchmarking to gauge the true progress of landcare in reversing the current trends.

The impact of the landcare movement on the long term economic viability of the farming community, rural adjustment and successful rural community cohesion needs intensive research. Additionally, the influence of landcare promotion on bridging the gap between rural and urban communities is also an important area for further research. Sustainable land management is the responsibility of the whole community (urban and rural); the success of the landcare movement is crucial if "Land, Water, and Life" is to be sustained for the benefit of future generations.

REFERENCES

ABARE. 1994. Survey of landcare and land management practices 1992-1993.

Australian Bureau of Agriculture Research Economics Bulletin.

Australian National Opinion Poll Survey. 1991.

Beale, B., and P. Fray. 1990. The Vanishing Continent. Hodder & Stoughton (Australia) Pty Ltd.

Campbell, A. 1994. Landcare: Communities Shaping the Land and the Future. Allen & Unwin Pty Ltd.

Chamala, S., and P.D. Mortiss. 1991. Working together for landcare. University of Queensland and Queensland Department of Primary Industries. Australian Academic Press, Brisbane.

Loton, B. 1995. Speech at national launch of Landcare Foundation at Institute of Company Directors Satellite Linkup Luncheon.

National Water Quality Strategy. 1992.

Prime Minister's Science Council Booklet. 1995.

Roberts, B. 1991. Lessons from the Australian landcare movement. Proceedings from the International Conference on Sustainable Land Management, Hawkes Bay Regional Council, Napier, New Zealand.

Scarsbrick, B. 1992. Corporate involvement in the landcare scheme. Business Council Bulletin December.

Shoebridge, N. 1992. Business Review Weekly, May.

12

Privatization of Extension in the United Kingdom: Implications for Agriculture and the Rural Economy

Michael Bunney

The privatization of extension in the U.K. was triggered by a combination of the pressure to reduce U.K. government funding support for agriculture and the recognition that policy objectives had changed. The rationale for the decision to start charging for advice to farmers was set out by the Director General for Agricultural Development and Advisory Service (ADAS), Professor Ronald Bell in his study report (MAFF 1984).

Charging for some services commenced in 1987 and the initial basis for this privatization process was the recovery of costs. The objective has subsequently become to transfer a complete section of the government advisory services into the private sector. This covers ADAS in England and Wales, and the Scottish Agricultural Colleges (SAC) in Scotland. Circumstances in Northern Ireland have restricted the process.

This transition from state-controlled, production-orientated policies towards market-driven, customer-orientated policies for the agricultural industry was driven by the basic premise that "those who benefit should pay." 'Advice to farmers' in this context included all those activities involved in information and technology development and transfer, and the on-farm advice provided to the industry and its individual businesses.

Further pressure to move the extension process into the private sector as soon as possible is coming from the requirement to reduce all levels of public expendi-

ISBN 1-57444-104-3/98/$0.00/$.50

ture in the U.K. A timetable has been set to privatize fully at least a portion of the existing state-controlled service. The criteria to be used for determining the boundaries of the split have yet to be agreed and announced. This debate goes to the heart of the future role for extension in agriculture and the rural economy.

The consequences of this policy change have yet to be fully demonstrated. In the context of technology and information transfer, ADAS has been able to demonstrate that there is a distinct market for certain types of advice and services to particular sectors of the industry. The continuing debate includes how far the process can be driven by the private sector and what impact this will have on the delivery of other government policy programs. The current position and the basis for the debate needs to be considered in the context of the policy programs that have operated over the last 50 years.

POST-WAR POLICIES

Post-war Britain faced a particularly challenging situation in relation to all economic development, but one of the highest priorities was set for the agriculture and food supply sector. The policies that were set out in the 1947 Agriculture Act were based on three main objectives.

The most critical was the need to be more self sufficient in all temperate food supplies. War-time experience had convinced all parties of the strategic significance of this, but there was also the wider consideration of trade imbalances at a time when other industrial output was at a low level. Second, it was recognized that increased food output could only be achieved by improving the technology employed in farming and by improving the efficiency of the production systems.

The third objective was to develop the farming business and marketing structures, to enable the efficiencies of production to be translated into the food chain. The food market itself was minimal, with the nation facing shortages, rationing, and seasonality of supplies. The supply infrastructure was basic, centered around local shops, local supplies and markets, and minimal logistics.

It was considered that such a major policy program to meet this national and public need for primary agricultural products could only be delivered from a state-controlled development and extension system. The structure reached right down to local districts within all the counties. In England and Wales this was an integral part of the Ministry of Agriculture, Fisheries and Food (MAFF). In Scotland the structure was based on the Scottish Agricultural Colleges of the three universities.

The difference between the English and Scottish systems was that in England, education and training was delivered by separate organizations from the advisory services; whereas in Scotland, the system was more akin to the U.S. land grant university approach, with research, education, training, technology transfer, and advice to farmers being delivered through the college structure. In both countries, agricultural research was also carried out by separate institutes as well as in the universities.

The extension process was a combination of demonstration of technologies, farming systems, and techniques (using MAFF Experimental Farms and Horticultural Stations, as well as commercial farms that had adopted the particular concepts); advisory campaigns to encourage adoption through discussion groups, conferences, agricultural shows and bulletins, and advisory leaflets; and one-to-one, on-farm advice direct to individual farmers covering the full range of technical and farm business issues (sometimes linked to grant aid such as schemes to encourage account record keeping or business planning).

POLICY ACHIEVEMENT AND THE NEED FOR CHANGE

This highly concentrated program resulted in rapidly expanding output of agricultural commodities from the late '50s onwards. A wide range of intensive technologies were adopted, and it was generally believed that the process was following the defined route from fundamental research through development to the advisory programs that stimulated uptake by the industry. This technology transfer process was underpinned by the grant-aided capital investment programs and the deficiency payments system that supported the main commodity prices.

The cost of these programs escalated rapidly from the 1960s onwards. Control of these costs was attempted first by fixing ceilings on the deficiency payments schemes for the main commodities. Grant aid schemes were geared more towards the efficiency of production, rather than increasing the volume of production. But the principles of state agricultural research and extension remained, as it was considered that this was still the most effective way to achieve technology adoption across the bulk of the industry.

Even when Britain joined the European Economic Community in 1971 and the price support mechanism for agricultural commodities changed from the deficiency payment system to intervention buying, the principle of extension continued.

ADAS was officially formed in 1971, bringing together all the various professional, scientific and technical services within MAFF under a single, more streamlined management structure. It included the National Agricultural Advisory Service (NAAS), which had its origins in the wartime County Agricultural Executive Committees—the 'War Ags'—and plowing up policies and other measures to help Britain counter the threat to its overseas food supplies. Other services merged into ADAS included the State Veterinary Service (SVS), the Agricultural

Land Service (ALS), the River and Coastal Engineers, the Drainage and Water Supply Officers, and the Horticultural Marketing Inspectorate (HMI).

Under the EEC Common Agricultural Policy, the cost of the agricultural support program to the U.K. Treasury continued to increase during the 1970s. By then, the U.K. was about 85 percent self-sufficient in temperate food supplies and the rapidly developing world trade system was enabling many seasonal products to become more available throughout the year. At the same time, sections of the public were becoming more concerned about the impact that intensive farming activities were having on the environment, wildlife, animal welfare, and human health.

By the early 1980s, the need to contain the agricultural support program was coinciding with the need to reduce overall government spending in the U.K.. Policies under question affected research, development, and the advisory services.

The National Economic Development Office, Agriculture Economic Development Committee, set up a project to study the uptake of technology in agriculture. Its report (NEDO 1985) recommended that improvements were needed in the procedures for determining the R&D program, the setting of priorities, and in the operation of the advisory services.

Several of the case studies on which the NEDO report was based, indicated that the process of technology adoption in the U.K. did not always follow the classic technology development and transfer chain defined by the state extension system. The process did not always meet market needs and the private sector sometimes met these needs more effectively than had been recognized.

The NEDO report went on to draw attention to the duplication of extension research, development, and promotion, which was already being provided by the private sector. The report also noted the potential for private sector involvement to be increased.

It also suggested that a more market-oriented approach be adopted to promote new technology, and that the advisory services should assess the usefulness of merchants, contractors, and consultants for promoting specific technologies. The report emphasized that farmers would need to be more self-reliant, would need to develop their business management skills, and would need to take more specific professional advice.

While studies such as the NEDO report were reviewing the technology transfer process in the primary production sector, it was also very apparent that the food market had also changed. There was increasing national wealth. More people were able to be less concerned about basic survival and security and to start to seek greater choice and quality in the products and services they bought. This, in turn, gave rise to the modern trend in supermarkets, dining out, and an expectation of year-round supply of formally seasonal foods. These trends stimulated a market for differentiated and competing products, with a sophisticated world-wide food chain to supply these demands.

A policy on extension that had served the country well in the post-war years was now considered to have outlived its requirements, as market forces were replacing the need for intervention. What had been necessary as a public good could now be regarded more as a private good with profitable market opportunities.

There had been a reluctance to make any significant changes in the 1970s, so the requirement became more acute by the mid-1980s. The Conservative Government that had been elected in 1979 with an agenda to deliver a more market oriented program felt more able to tackle the issue after its reelection in 1983.

ADAS AT THE CROSSROADS

"Turning cozy public monopolies into fighting fit responsive operations is a pretty traumatic and painful process," was a statement made frequently by the marketing consultants brought in by MAFF and ADAS to advise on the commercialization process that started in 1985 (Drakes 1986).

In Britain in 1985, state controlled organizations still dominated the technology and information transfer process in the agricultural industry. The Agriculture and Food Research Council (AFRC) controlled the research institutes responsible for the more fundamental research. ADAS carried out its development role on this primary research through the experimental husbandry farms and horticultural stations, along with trials on commercial farms. It also included the Central Veterinary Laboratory (CVL) and the Central Science Laboratory (CSL) within its resources.

The advisory services were structured by discipline, such as agronomy, dairy husbandry, mechanization, and microbiology. General agriculturists provided the front-line first point of contact for farmers.

Education and training were and still are the responsibility of the universities, colleges, and a training board, though structures and designations have changed. As stated earlier, the pattern in Scotland was different, with the advisory services linked through the universities, but still with similar concepts. The commercialization process was applied to Scotland at the same time.

THE COMMERCIALIZATION OF ADAS

The marketing and business development process that was initiated in 1985 included all the elements normally associated with commercial development. The clear objective was to determine what market there might be for the types of services, such as direct advice to farmers, that had been provided by the state, and to test what other types of services might be initiated. The only restrictions were in the areas of environment and conservation, farm and rural diversification, and animal welfare, where advice would continue to be provided free by ADAS—a continuing extension role on behalf of MAFF.

The initial financial targets were in terms of recovery of the costs of running the organization. In 1985 ADAS was an integral part of MAFF, with an organizational structure that matched the vertical framework of MAFF from a headquarters in London, regional and divisional centers, and county-based staff within districts.

All ADAS accounting systems were linked into MAFF and the standard Public Expenditure Survey (PES) Vote Codes. Nobody knew what it cost to deliver particular services or to run particular parts of the organization. The total running costs of ADAS were estimated at around £100m (1 $ U.S. = 0.57 British pounds at time of publication).

Creating a commercially viable body out of this situation was not going to be achieved overnight. It was felt necessary to allow for two years preparation before any charging for services should begin. During this time an extensive program of market research, internal assessments, product development, staff training, and structural reorganization took place. (Bunney and Bawcutt 1991; Houseman 1993).

The process of creating a leaner and more market responsive organization was started and has continued by stages up to the present day. Today ADAS is a separate agency, still owned by MAFF and the Welsh office, but with a separate headquarters now in Oxford. The structure has been considerably flattened, with day to day management delivered through the single level of business centers (managing the consultancy services) or research centers (managing R&D work).

Staff now operate through business teams that are market sector oriented (such as dairy teams that include a mix of the disciplines involved in giving advice to that sector), rather than grouping staff by discipline to service all industry sectors (such as the microbiologists' discipline that covered all aspects of livestock and arable farming, and on into the food sector).

ADAS is aiming to operate from around only 20 offices compared with the 80 it occupied in 1992. Many consultants now work from home using modern information management systems for communication and administration.

Several parts of the original ADAS have been split off or retained by MAFF, including the SVS, CVL, CSL, and HMI. The Experimental Horticultural Stations have been transferred to Horticulture Research International (HRI). At the same time there has been an overall reduction in staff as market demand has been assessed and new working practices introduced. Today ADAS has around 2,200 staff compared with around 6,300 in the early 1970s.

CUSTOMER FOCUS

The strong feature that has enabled ADAS to create a significant market for consultancy was the decision back in 1985 to base a marketing strategy on developing products and services related to market demand and customer needs, rather than try to put a price on the old-style extension services.

Previously, the private sector consultancy services were very limited and mainly concentrated in the arable sector. The main exception was the private veterinary sector, but ADAS had never been involved in this context as the SVS was responsible for government animal health regulations and required a clear distinction between roles. The other private consultants were mainly involved with pesticide and related advice in crop management, with some private advice on business management issues.

The main competition to ADAS was seen as the services supplied by the merchant representatives of the main agricultural supply companies. Their ‘free’ advice was shown to be a major source of day to day information to farmers, but was tied closely to specific products.

The agricultural supply industry saw a range of opportunities arising from the commercialization process. It enabled them to refine the services they supplied to farmers, as all advice was now seen to contain some commercial element. They were also able to buy advice from ADAS and to use ADAS to carry out research for them, particularly comparative trials on products. This led to some difficult relationships on product endorsement until better agreements had been established.

Private consultants also started to extend their areas of activity and to feel able to compete on more equal terms—though there were still strong complaints that ADAS was receiving subsidies that were not available to the private sector. In practice, because ADAS had to recover its full costs—‘Full Economic Costs’ (FEC)—on all work done and had to carry higher overheads to service its government commitments, the fixed daily rates were higher than many in the private sector, particularly for technical advice. The ‘subsidy’ covered by MAFF made up the annual shortfall, but Treasury targets set increasingly tough requirements to reduce this each year, eliminating it by the end of the 1996/97 financial year.

ACHIEVEMENTS OF THE NEW ADAS

ADAS was able to create a new market for commercial consultancy that has grown to around £20m per year (Table 1). Other players have come into the market, but at the same time ADAS has been able to exploit its expertise in new market areas that were not available to it while it was within the MAFF structure by applying existing expertise into new contexts.

Table 1. Consultancy Revenue Growth (1987-1995)

Year	ADAS Consultancy Revenue (£m) (1 $ U.S. = 0.57 £m)
87/88	6
88/89	8.7
89/90	12.1
90/91	15.1
91/92	16.4
92/93	18
93/94	19.3

These new market areas have been loosely classified as the land-based industries, but extend into a wide number of business activities that have land as one of their basic resources. These include a wide range of environmental and land reclamation services (such as opencast mining and quarry restoration); alternative land use activities, particularly sport and leisure (such as golf course construction); and diversification into new crops for energy production and other industrial purposes. ADAS has also been able to extend into the food service system, particularly in relation to food hygiene and quality management systems, from its strong base in microbiology and food technology.

At the same time, ADAS has been able to build up a strong market for private research and development work, beyond that which it carries out directly for government. Some of this is in competition with private research companies for the levy (check off) funds raised by the statutory bodies established to commission R&D work on behalf of particular sectors of the industry such as cereals, milk, and horticulture. Other R&D is directly for commercial companies, crop and livestock centers, and private groups of farmers. This area of work has reached around £6m revenue per year, despite the loss of some facilities (Table 2).

Table 2. R&D Revenue Growth (1990-1995)

Year	ADAS R&D Revenue (£m) (1 $ U.S. = 0.57 £m)
90/91	1.7
91/92	4.1
92/93	3.9
93/94	4.5
94/95	4.9

This strong research base, involved with both private research as well as research for government, has enabled ADAS to demonstrate its credibility and independence in the quality of advice being offered to the industry. It is on this basis that it has set out its mission statement "to be the leading consultancy to land-based industries in the U.K., working with our customers through the provision of quality services for the benefit of their businesses."

Over the last three years, overall revenue to ADAS has been relatively static (at around £75m). This reflects the declining amount of work for government, even though commercial revenue has increased. Total costs have been declining steadily, so that the deficit between revenue and costs has been reduced to around £10m (Table 3). Part of this cost saving has come through reduction in staff numbers (Table 4), but significant savings have been made through the rationalization of offices and working practices.

A significant outcome of the commercialization process has been the development of effective management information systems, to enable management to identify the true costs and returns of all ADAS activities and to relate these to their business objectives and plans.

Table 3. Cost Recovery Achievements (1992-1995)

Year	Total Revenue (£m)	Total Costs (£m)	Net Cost (£m)
92/93	75.82	93.1	17.28
93/94	76.02	90.18	14.16
94/95	75.72	85.99	10.27

Table 4. ADAS Staff (1991-1995)

Year	Staff Numbers
91/92	2900
92/93	2600
93/94	2555
94/95	2195

PRIVATE VERSUS GOVERNMENT FUNCTIONS

While the emphasis of this paper has been on the commercialization process, it should not be forgotten that two thirds of ADAS work in the 94/95 financial year was still on behalf of government, either as R&D work (24.2 percent) or in running extension programs for MAFF and other government departments (42.3 percent) (Table 5).

ADAS is 'paid' by MAFF at the FEC rate for each of the staff-years employed on these programs, known as Memorandum of Understanding contracts (MOUs). One of its key performance targets for its owners—MAFF and the Welsh Office—is a regular annual reduction of the total cost per person employed, which is the basis for this FEC figure.

A parliamentary statement by government in June 1995 confirmed that a number of ADAS' functions are suitable for privatization. Provided that it meets the target of recovering 100 percent of its costs for advisory services through charges to its commercial customers, the target date of 1997 was set for a sell-off. A senior management team has already indicated its interest in setting up a management

Table 5. ADAS Revenue from Main Activities

Activity	Revenue Breakdown (%) 94/95
MAFF Consultancy	42.3
MAFF R&D	24.2
Commercial Consultancy	25.6
Commercial R&D	7.9

and employee buy-out and is working hard to achieve a profitable position by that date.

The functions likely to be sold off will include all the commercial consultancy services and the laboratory services. It will also include the R&D services, though no decision has yet been made over the ownership of the land and facilities.

Certain non-R&D functions for government, mainly in connection with the agri-environment schemes, are likely to remain in the public sector where bias is an issue. The precise details of this split have yet to be announced, but they will have to be determined by the end of March 1996 if the timetable is to be achieved.

IMPLICATIONS OF PRIVATIZATION

There has been only limited public debate on what functions should be retained in the public sector. Government argues that it is possible to deliver many of the activities required to implement policy through the use of private agencies, by contracting out services and by 'Market Testing'—whereby existing public services have to compete against outside bids for the delivery of their current services. As a result the debate has concentrated more on the method and funding of delivery and less on what issues should involve government action.

At present there seems to be a line being drawn between the more sensitive public good issues that will continue to be handled within the public sector services and those functions that can be delivered equally effectively by private services but with taxpayer savings.

The market driven philosophy that has operated over the last 18 years in the U.K. is based on the premise that cost effectiveness is best demonstrated by private sector market forces. The demand for reductions in public spending will continue to put pressure on ensuring that the cost of delivering public services to the industry will be strictly controlled and alternative methods of more cost effective delivery will be explored.

The other side of the debate that has not had such wide discussion relates to the future role of government in intervening in and supporting the agricultural and rural economies. There is an assumption that interference in the marketplace should only take place where clear market failure problems are arising. When allied with the pressure to reduce government expenditure and, more specifically, to reduce spending on the U.K. agriculture sector, the case for any intervention must be strongly argued.

Until the decision is made on where the split in ADAS will be made, it is difficult to assess the implications for the priorities for the role of agricultural extension in the public sector. The commercialization process of ADAS has demonstrated that as the agricultural industry becomes more dependent on the market and less on state intervention, so it is able to gain many of its main technology and information needs from services in the marketplace.

This demonstrates that many areas of advisory work that used to be considered a public good can now be classified as a private good, where a profitable market can be established. This relates particularly to the one-to-one, on-farm advice that has direct benefits to the farmer.

But the process has also shown that only a proportion of farming businesses use these services to any significant amount. While there has been no published evidence, experience of working in the new structure suggests that perhaps up to 80 percent are not using them to any great degree. While the other 20 percent may be responsible for a high proportion of agricultural output, the 80 percent represent many small family farms in the more rural and less developed parts of the country.

There is no evidence that this was an intended outcome of the commercialization process, or that it was intended to speed up any restructuring of the industry into larger units. Instead, in the early stages of the commercialization process, ADAS was encouraged by government to offer special services to meet the needs of the smaller businesses. A fixed, low-price subscription scheme offered a combination of newsletters, telephone advice, and one farm visit. It was also suggested that small groups of farmers could buy consultants' time on a collective basis, but only be charged at the going rate for one-to-one consultation.

These offers were not successful, though there was greater uptake in the more remote and extensive farming areas, such as Wales. They were difficult to promote, often conflicting with the promotions by consultants concentrating on their mainstream activities and by the readiness of farmers to accept the agricultural supply trade advice (despite potential bias towards particular commercial products and systems).

Consideration is now being given to meeting their needs through the broader services targeted at small businesses in general, such as those being developed through the Department of Trade and Industry's Business Links. Business Links are being established to bring together services and support from all government departments as a local 'one-stop-shop' for all industry inquiries.

At present such services are not capable of delivering the more sector-specific services and advice needed by farming businesses, as this has been the role of MAFF and ADAS. But one business link is recruiting an agricultural business adviser to assist with programs for that sector. MAFF is conducting a pilot project to establish how best Business Links and MAFF can provide a service by which each can 'signpost' each other's services to rural businesses (DoE, MAFF 1995).

The residual ADAS body, to be retained in the public sector, may be designated a role in this context, but this also awaits the government statement on the split.

Further evidence of the requirements for business advice and training for small businesses in rural areas comes from a report based on a study highlighting the different needs between rural and urban areas (NRETS 1996). These areas differ in terms of socio-economic infrastructure, environmental characteristics, and physical distances. Businesses tend to be smaller, with more self-employment, and net emigration of young people in contrast to immigration of older people. There also appears to be even more lack of business and management skills in rural areas. These differences suggest the need for more specific government action and points to an extension process as the appropriate method.

The large-scale farming businesses will also require effective sources of independent advice. The trends in the food chain are for more vertical integration, drawing groups of farmers into production systems geared to particular product lines for supermarkets or other outlets. As biotechnology and other developments evolve, so may this vertical linkage become even more significant. Farmers will be making decisions to enter into particular alliances, often from weak negotiating positions, due to poor development of their own business and marketing structures. In the past, advice and support for such developments was available from government services.

The network of intelligence provided by the agricultural extension process was able to address a wide range of issues, beyond the immediate technical and business concerns of the industry. It encompassed a wide spectrum of individuals from the academic and research based through the advisory and support organizations and into the business and ground level activities. It is not yet proven that the market driven approach, with its more self-centered, short-term monetary attitudes, will be capable of taking account of the wider long term issues such as the sustainability of farming systems.

Government itself will continue to require sound professional policy advice. Up to now much of this advice in the agriculture sector has been available 'in-house', supplemented by external sources as required. As more and more of these professional elements are moved into the private sector, government will seek more external advice for its policy development.

The remaining senior civil servants responsible for developing policy papers for ministers will be 'generalists' relying on sound professional advice for the basis of their submissions. This system has yet to be tested as to how far this process can proceed before difficulties arise in providing reliable independent advice to ministers.

The present aim is to enable the market to meet many of the needs of industry. It is recognized that government must intervene where there is market failure or where national and public needs outweigh the interests of the particular sector. As more of the extension services are moved into the private sector, it is becoming less clear whether they will be able to identify and advise on market failure in the same manner as previously.

This will be the case particularly if these services start to lose their independence and become more closely linked in one way or another with particular commercial interests. In the same way that groups of farmers may find business pressures encouraging them to join alliances, so similar pressures may impinge on independent private consultancies.

The role of universities and private policy units may become more important in this context as a means of researching issues. There may still be elements of bias creeping into some of these inputs, which suggests that further mechanisms will be required to ensure the arguments are fully considered through open consultations before actions are taken. A greater reliance on a wider range of external policy advice, with less in-house input, may enhance the overall process of policy development and the setting of priorities. This approach has been less tested in the agricultural sector compared with other industries.

CONCLUSIONS

The commercialization process of ADAS has shown that it is possible to change the objectives and priorities for agricultural extension and enable a more market driven approach to meet the mainstream technology and information needs of the leading farming businesses.

The process has also demonstrated that it is possible to convert a public sector extension service into a market driven consultancy, to meet the needs of the industry, and to diversify into new areas of business. The expectation is that the research and commercial consultancy arm will be converted successfully in the near future into a privatized business, outside the control of government.

One of the significant benefits of the commercialization process has been the introduction of high quality management information systems. This has not only enabled the true costs of all the extension activities to be identified, but also enabled the organization to demonstrate improved 'value for money' from its services to government. Costs of services can be broken down to constituent parts, with the outcomes of activities being measured against the original objectives and the relevant costs of delivery.

Public extension services will still be required to meet the wider national and public needs that cannot be addressed by market forces. Government will still intervene where market failure occurs or where national needs override industry sector needs. The priorities will relate to the agri-environmental issues and the rural economy/small business development needs. Extension, with its education and training role, will be particularly important in the latter context.

The process of definition and identification of market failure will alter as result of all the organizational changes. It is not clear how this process will be met in future or whether the present drive for reduced public expenditure and withdrawal from a wide range of interventions will override or limit the identification process to the detriment of the longer term development of the industry.

Policy program implementation for a range of issues is still considered best served by publicly controlled extension services, but with a choice of delivery mechanisms ranging from 'in-house' government departmental services, 'arms-length' government executive agencies, short-term contract services, or the employment of private companies to deliver complete programs.

The implications of the full privatization process have yet to be assessed. Delivery mechanisms for government programs by private sector arrangements will continue to be tested against the traditional use of the government's own services. The way that market failures and wider public good issues will be recognized and addressed will determine the scope for future extension programs.

With the dismantling of existing agricultural extension and advisory services, there will be more limitations to the availability and access to policy advice and for providing future services. This, in turn, may limit the rate of response and uptake of new technologies and information across the industry and so reduce competitiveness in an industry that is still heavily dominated by relatively small businesses and weak marketing structures.

REFERENCES

Bunney, F.M.G., and Bawcutt, D.E. 1991. Making a business of an extension service. Agricultural Progress - the Journal of the Agricultural Education Association. 66:36-43.

DoE, MAFF. 1995. Rural England: A nation committed to a living countryside. Government White Paper, Cm. 3016. Department of Environment and the Ministry of Agriculture, Fisheries and Food. HMSO. London.

Drakes, D. 1986. Getting to first marketing base. A presentation to ADAS. Unpublished.

Houseman, C.I. 1993. A Symposium—Getting value for money from extension. Paper presented at the Australian Pig Science Association Conference, Canberra, Australia.

MAFF. 1984. Report of a study of ADAS by its Director General, Professor Ronald L Bell.

Ministry of Agriculture, Fisheries and Food, London.

NEDO. 1985. The adoption of technology in agriculture: Opportunities for improvement. National Economic Development Office, Agriculture Economic Development Committee, London.

NRETS. 1996. Business advice and training in rural areas : Access and coherence. National Rural Education and Training Strategy Group, ATB-Landbase, National Agricultural Center, Warwickshire, England.

Section Four

INSTITUTIONAL INNOVATION

13

Technology and Information Transfer in U.S. Agriculture: The Role of Land Grant Universities

Nicholas Kalaitzandonakes and J. Bruce Bullock

Important changes have recently occurred in the process by which technology and information are transferred to U.S. agriculture. Private management consultants and industry sales representatives have become increasingly important in providing information to farmers (Ortmann et al. 1993; Wolf 1995). Much of the technical knowledge traditionally transferred to agricultural firms by land grant universities (LGUs) through the Cooperative Extension Service has become part of "technology packages" offered by input suppliers, agricultural cooperatives, independent consultants, and agricultural entrepreneurs. Contract production and precision farming are examples of such technology packages.

The changing relative importance of private and public sectors in agricultural technology and information transfer are largely motivated by technical and organizational innovation (Rhodes 1993; Stein 1995) as well as significant structural change in the agricultural sector. It is unclear, however, how such changes are impacting the organizational structure of LGUs as sources of innovation and technology transfer. A primary difficulty in assessing such impacts is the lack of a satisfactory conceptual framework that ties the structure of the agricultural sector to that of the public innovation system.

Helpful comments from Bruce Bjornson, Daryl Hobbs, Kitty Smith, and Steven Wolf are acknowledged.

ISBN 1-57444-104-3/98/$0.00/$.50

For many years, the dominant paradigm of agricultural innovation and technology transfer has been the linear rational model which assumes that if the scope for economically beneficial technology transfer exists, such transfer will in fact occur. This assumption implies first, that the research and technology transfer system is always aware of the end-users' technical needs and, science permitting, will always generate technologies that address them; and second, that end-users always understand the benefits of technologies generated by the technology transfer system and, given enough time and capital, will eventually adopt them. The fundamental implication then is that more science leads to more innovation and technical progress, all taking place in a linear fashion.

This paradigm of technology transfer has followed the Solovian tradition of exogenous technical change. As a result, the research and technology delivery system is exogenized in the linear model of technology transfer. Insistence on this theoretical model is exemplified by the numerous technology adoption and diffusion studies carried out by agricultural economists and rural sociologists over the last forty years. In these studies, the technology transfer process is reduced to a one way causality where "good" managers (e.g., educated, experienced) are expected to adopt new technology faster. The appropriateness of the technology generated and transferred is rarely questioned and the modes of delivery are typically obscured. In effect, the *linkages* between the research and technology transfer system and the end-user are ignored. Thus, there is little theoretical basis for judging how structural adjustments in agriculture may affect the research and technology transfer system.

In this paper we introduce a model of technology transfer which allows the dynamics of technology generation, delivery, and adoption to be studied within a unified framework. Within the proposed framework, the structure of the technology and information transfer organization can be tied to the characteristics of the end-user. Subsequently, we use the suggested model to examine how land grant universities are responding to structural change in U.S. agriculture and the privatization of technology and information transfer. Within the same context we also examine their likely future role in agricultural innovation and transfer.

THE LINEAR AGRICULTURAL TECHNOLOGY TRANSFER MODEL

Introduction of any new agricultural technology is typically met with only partial success. Many members of a population of potential adopters adopt more slowly than others while some never adopt. Such behavior gives rise to innovation-diffusion which typically follows rather standard sigmoid temporal patterns.

Due to such regularities, innovation adoption and diffusion studies in agriculture have dominated the literature focusing on comparisons between adopters and non-adopters in an attempt to identify systematic differences in economic, social, demographic, institutional, and locational characteristics of each group (Rogers 1983; Rogers and Shoemaker 1971; Feder et al. 1985).

The conventional wisdom of these studies is that constraints to rapid innovation adoption involve factors such as lack of credit, limited access to information, aversion to risk, inadequate firm size, insufficient human capital, absence of equipment to relieve labor shortages, irregular supply of complementary inputs, and inappropriate infrastructure (Feder et al. 1985). Implicit to such adoption and diffusion models is the assumption of exogenous innovation and technology generation. No direct influence is exerted by the recipient of the technology on the innovation process. The research establishment is assumed to add to the stock of scientific and technical knowledge which, presumably, is always appropriate to the needs of the potential end-users. Hence, the rational economic agent should unquestionably adopt the innovation in the absence of exogenous constraints. With such pro-innovation bias, the emphasis is clearly on the end-user (Abramson 1991).

The assumption of exogenously generated agricultural technology has been rejected in several studies. Huffman and Miranowski (1981), and Evenson and Rose-Ackerman (1985) provided empirical evidence where farm interest groups were found to exert influence on the funding and direction of agricultural research and technology transfer. Similarly, Busch and Lacey (1983) concluded that granting agencies, agribusiness firms, foundations and other organizations exercise significant influence on the direction of public agricultural research primarily through grants, contracts and consulting funds. In addition to being misleading, the assumption of exogenous innovation also conceals the linkages of the agricultural research and technology transfer system with its end-users. As a result, the bulk of the agricultural innovation and technology transfer literature provides little guidance about how the innovation and technology transfer system may change in response to structural changes in the agricultural sector.

A non-linear agricultural innovation model would be a step in the right direction for analyzing interdependencies between the technology transfer system and the agricultural sector. Non-linear innovation models have been developed to capture observed simultaneous influences of demand-pull and science-push forces and other complex selection mechanisms on the direction of innovation and technology transfer (Kline 1991). By providing specific directional schemata of innovation and technology transfer, non-linear models illustrate how particular structural characteristics of both recipient and the transfer organization act to influence such directions.

An important limitation of non-linear models is that they are structured around empirical findings and stylized facts. As such, they are descriptive rather than predictive models. They can rationalize observed behavior but provide little guidance on how the technology transfer organization or the end-user respond to change. In effect, non-linear innovation and transfer models lack a theory of human behavior.

Existing non-linear models of innovation and technology transfer also fail to encapsulate the tacit dimension of innovation and technology transfer (Senker 1995). As empirical evidence is mounting that the tacit dimension of technology is a primary determinant of success in technology transfer, it becomes clear that such considerations should be explicitly incorporated in innovation and technology transfer models (Pavitt 1991).

A BEHAVIORAL MODEL OF TECHNOLOGY TRANSFER

In this study we propose that technology transfer be viewed as a multitude of inter-connected exchanges of technical knowledge. Each exchange occurs between two parties: the "source" and the "recipient."[1] The source transacts to access existing knowledge, produces new knowledge through a creative transformation process combining existing knowledge and conventional inputs (e.g., labor and capital), and transacts to transfer new knowledge. The recipient transacts to access the transferred knowledge.

The knowledge-adding activity of the source may involve creation of entirely new knowledge of the natural world; application of existing knowledge for the creation of physical devices that solve specific technical problems, replication of existing knowledge for the reduction of error, inconsistencies, and other "noise", and synthesis and standardization of existing knowledge into new system designs. Thus, the knowledge-adding activity may comprise net additions to the stock of knowledge or, in Boehlje's (1994) terminology, refinement and transformation of knowledge into information.

The transfer of new knowledge from the source to the recipient has friction, and it is constrained by information shortages because the two parties have different information. The source is typically more informed about the new technical knowledge being exchanged while the recipient knows more about personal idiosyncratic position and effective demand for new knowledge. An important contributing factor to the differences in the information sets of the source and the recipient is the tacit dimension of knowledge. Polanyi (1966) noted that humans know more than they articulate. In most cases, knowledge accumulated through experience and practice is not formalized and codified. Such knowledge is not accessible by everyone, which leads to informational asymmetries. Asymmetries in information may also be the result of uneven effort and unequal resources devoted to gathering knowledge by the two transacting parties.

For informational asymmetries to be reduced and knowledge transfer to occur, the source must transact to evaluate the demand for new knowledge and locate potential recipients while the recipients must transact to assess the relevance and potential impacts of new knowledge or seek new knowledge appropriate to their needs. The basic implication here is that because of informational asymmetries, the amount of new knowledge *added* by the source is not necessarily equal to the amount of new knowledge *transferred* from the source to the recipient. The proportion of knowledge added that is transferred is thus dependent on the resources devoted by the source and the recipient in transacting.

The proportion of knowledge added that is effectively transferred is also dependent on the conditions of transfer. A variety of modes and mechanisms can be used to transfer knowledge with varying degrees of effectiveness. They include: passive transfer of knowledge (e.g., technical reports or journal articles), active transfer of knowledge (e.g., workshops and seminars), apprenticeship, collaborative work, and others. The relative effectiveness of these mechanisms depends on their efficiency to facilitate the way in which people communicate, work together, and learn from each other. The more interactive and participatory the mode of knowledge exchange is, the higher the proportion of knowledge transferred will be.

The transformation and the transaction processes require resources and hence they are costly. Transformation costs incurred by the source involve the costs of capital, labor, and other conventional inputs that are combined with existing knowledge to create new knowledge. Higher transformation costs imply use of larger amounts of inputs which lead to higher expected knowledge added.

Transaction costs are incurred by both transacting parties in the exchange of new knowledge. Transaction costs incurred by both the source and the recipient involve expenditures associated with reducing informational asymmetries. Additional transaction costs are incurred by the two transacting parties for monitoring and enforcing agreements and property rights in the transfer. These later transaction costs reflect the possibility of opportunism and misinformation deliberately provided by any one of the transacting parties when it is profitable (Williamson 1989).

Transaction costs tend to reflect the idiosyncratic conditions of exchange and non-tangible factors such as trust, willingness to cooperate, and ease of interface. As a result, transaction costs in technology transfer increase with the degree of knowledge tacitness and the complexity and uncertainty in the knowledge exchange; while they decrease with the frequency of exchange and the strength of relevant property rights.

Both the source and the recipient are assumed to maximize net benefits from their exchange. Such benefits are the difference between gross benefits and the sum of transformation and transaction costs. The benefits accruing to the recipient are determined by the value-in-use of the transferred knowledge. For the source, the benefits are tied to the derived demand for new knowledge.

The optimal level of knowledge transferred from the source to the recipient is determined jointly by the two transacting parties. Transferred knowledge is thus dependent on benefit structures and transformation costs but also on transaction costs which vary with the idiosyncratic conditions of the source, the recipient, and their exchange. Under these conditions, the amount of knowledge transferred may be explicitly tied to demand-pull and science-push forces as well as other selection mechanisms that tend to impact the transaction costs of technology transfer. For example, transaction costs may explain the effectiveness of technology transfer programs shaped by demand-pull rather than science-push forces. When technical needs are brought forward by potential recipients the source need not expend scarce resources to identify such needs or locate recipients interested in the transfer. Similarly, recipients need not expend resources to understand the relevance of the new technical knowledge. Under such conditions, information asymmetries between the recipient and source are reduced at low transaction costs and transferred knowledge tends to increase.

To summarize, the key elements of the proposed behavioral model of technology transfer are as follows:

Human behavior is assumed to be consistent with standard microeconomic theory. In particular, both transacting parties maximize net benefits subject to a set of relevant constraints.

The tacit nature of knowledge is explicitly incorporated. The higher the tacitness, the larger the associated transaction costs will be, reflecting the difficulty of transferring such knowledge and the costliness of engaging in active delivery modes to increase the level of knowledge transferred.

Non-linearities in innovation and technology transfer are allowed for by differentiating between knowledge added and knowledge transferred, and by allowing demand/supply forces as well as the conditions of transfer to determine the proportion of knowledge added that is effectively transferred.

Structural characteristics of the recipient and the source are integrated through their influence on transaction costs and hence on the level and direction of technology transfer.

AGRICULTURAL TECHNOLOGY TRANSFER AND TRANSACTION COSTS

The proposed model may be used to analyze the behavior of LGUs, which are viewed here as sources of agricultural innovation and technology transfer. In the previous section, we argued that transaction costs are a portion of the overall costs incurred in technology transfer. In this section we will argue that due to specific attributes of technology transfer in agriculture, transaction costs incurred by LGUs and some potential recipients are likely to be large. As such, they could have significant impacts on the demand for knowledge transfer as well as the behavior and organizational structure of LGUs.

Transaction costs incurred by farmers in agricultural technology transfer are augmented by inherent difficulties in accurately measuring the performance of new knowledge applied in agricultural production. For biological systems, it is often difficult to decouple the contribution of a particular technology from external random influences. For example, the ability of a farmer to assess whether poor seed performance is due to inferior genetics or bad weather is often limited. Decoupling and measuring the contribution of any single agricultural technology is difficult to assess not only because of random natural influences but also because of its connectedness with other factors contributing to increased efficiencies. For example, it is difficult to decouple productivity contributions due to advances in animal genetics from improvements in nutrition and animal health or from "management factors." Under such circumstances, the in-use value of and demand for such new knowledge decrease since significant transaction costs must be incurred to closely monitor and assess performance.

Transaction costs incurred by LGUs in agricultural technology transfer tend to increase primarily from the need to coordinate with a large number of geographically dispersed recipients with diverse structures and needs. LGUs contribute new knowledge to a diverse set of potential recipients including farmers, government bureaucrats, agribusiness managers, and other decision makers. Within any one particular group of potential recipients differences can be pervasive. For example, variations in agroclimatic conditions, firm size, production systems, capital, and management structures imply that the needs for new technical knowledge among farmers are very diverse. For LGUs to develop representative knowledge about the technical needs and effective demand for new technical knowledge of all potential recipients, substantial transaction costs are involved. Interacting and coordinating with a large number of diverse and geographically dispersed recipients to transfer new knowledge further increases transaction costs.[2]

Transaction costs incurred by LGUs in technology transfer have not been previously measured and have attracted only limited attention in rent seeking activities for agricultural research (Huffman and Just 1995). Thus, it is not known how large such costs can be. Some empirical evidence that transaction costs incurred by LGUs in agricultural technology transfer are large is provided here through a case study. Using survey information we measure transaction costs in knowledge transfer activities as the time of agricultural scientists spent transacting for accessing and transferring knowledge. We find that over half of the total scientist's time is, on average, devoted to transacting (see Appendix A for details). Thus, the proposition that transaction costs incurred by LGUs in technology transfer are large is supported by the case study.

In addition to being large, transaction costs incurred by LGUs are likely to be more malleable than transformation costs. Many of the essential inputs involved in the transformation process are typically lumpy and fixed. For example, the human capital involved in the creative process of adding value to existing knowledge is typically highly specialized and can not be easily shifted from one transformation activity to another. The same is true for specialized capital equipment. As a result, adjustment in the transformation process may be difficult, at least in the short run.

If the transaction costs are in fact large and more malleable than transformation costs, they will tend to dominate the behavior of LGUs and the direction of its technology transfer, especially in the face of change. In this study we focus on transaction costs to analyze the behavior of LGUs and their response to recent structural changes in U.S. agriculture.

LAND GRANT UNIVERSITIES AND STRUCTURAL CHANGE IN U.S. AGRICULTURE

Over the last several decades, agricultural production has become increasingly science-based. As technology in agricultural production has become progressively more sophisticated, farmers have been forced to search for increasingly complex and diffuse information on technology and business strategies. At the same time, the ability of the traditional extension agent to keep up with the ever-expanding information base has been limited. As a result, the traditional delivery mode of LGUs has gradually become obsolete, in part due to its own success in transforming agriculture.

The reduced role of extension and the increased needs for information by farmers have increased opportunities for private technology transfer. Most opportunities have been created in the provision of specialized technical knowledge and services. Increasingly, such knowledge is integrated with inputs and services into *packages* that reduce transaction costs incurred by the farmer. For example, sales of fertilizer and seed are packaged with soil analysis services and on-site expert advise. Similarly, improved genetics, feed rations, building technology, and on-site consulting have been packaged in hog contracts that have become increasingly popular (Rhodes 1993). All such technology packages tend to reduce search and coordination costs (e.g., costs of collecting information on individual technology performance) incurred by farmers. They also reduce measurement and monitoring costs since it is far easier to monitor and measure performance of a total package rather than a set of complex individual components whose performance is difficult to isolate. As agricultural production continues to become more science-based and complex, the demand for technology packages is likely to increase because of decrease transaction costs and hence increased net benefits to the farmer.

Despite the growing demand for diverse multi-disciplinary knowledge and technology packages, LGUs have not been able to service such demand. They have continued to excel at knowledge-adding activities where basic scientific knowledge is generated, but have been unable to integrate such knowledge into decision-ready transferable form for an increasingly diverse agricultural sector. Such tendencies have been encouraged both by large transaction costs in the transfer of knowledge and a changing incentive structure.

Agricultural scientists have been receiving weakening demand-pull signals as their traditional link with the farming sector—the extension agent—has been breaking down. Under such circumstances, the search and coordination costs for establishing and maintaining communication channels with clientele increased substantially. Search costs for setting research priorities are far lower for disciplinary and basic research where priorities are better understood by individual scientists, and so there has been a turn toward knowledge-adding activities and basic research.

Over the same period, the number of agricultural holdings have continued to shrink and become more commercial. The agricultural sector has progressively lost its "uniqueness" and so have LGUs that service it. As a result, the contributions of agricultural scientists in LGUs have been increasingly measured with disciplinary standards. These developments further re-directed the attention of agricultural scientists away from applied research and transfer, and towards disciplinary research conducive to publication in disciplinary journals and grantsmanship.

Limited empirical evidence supporting our proposition that agricultural scientists in LGUs tend to minimize contact with client groups so that they minimize transaction costs and maximize productivity in disciplinary research is provided in this study through our case study detailed in Appendix B. Our results indicate that emphasis in disciplinary research is not only motivated by minimization of transaction costs, but also by an incentive system that is perceived to reward such research.

While the emphasis in disciplinary knowledge generation may result in lower transaction costs, it may also result in lower value for the new knowledge produced by LGUs. Boehlje (1994) has argued that information (i.e., context-specific and decision-focused knowledge) is more valuable than general knowledge. This differentiation in the value of general knowledge and information may be valid only in a static sense.[3] It does suggest however that disciplinary knowledge, which is not decision-focused, may have lower value-in-use, at least in the short run.

Whatever the value of new knowledge generated by LGUs may be, it has become progressively more difficult to demonstrate within their traditional client group, the farming sector. Complexity of science and connectedness of public with private R&D and technology transfer activities obscure the contribution of knowledge generated by the public agricultural research system. Such difficulties have intensified as public technology transfer has been progressively "dehumanized" by the gradual withdrawal of the extension agent. Lower perceived value of knowledge is thus likely to result in diminishing demand for LGUs' knowledge output transferred to production agriculture through traditional vectors and methods.

THE FUTURE ROLE OF LAND GRANT UNIVERSITIES IN TECHNOLOGY AND INFORMATION TRANSFER

Decision making in agriculture, both for private firms and public regulators, will continue to increase in complexity. Environmental regulation and compliance, food safety, vertically and horizontally coordinated governance structures, introduction of sophisticated information and biotechnologies, globalization of food markets and quality standards, and changing weather patterns are some of the factors that will determine the complexity of decision making for years to come. With the realization that such complex issues cannot be addressed in isolation, the demand for diverse, system-wide information and technologies that support decision making will also increase. Hence, multidisciplinary information and technical knowledge packages that reduce transaction costs of the recipients, will be increasingly valued.

Opportunities for technology and knowledge packages which assist decision making in production agriculture will continue to expand for the private sector. Most of these packages will be facilitated by advanced technologies (e.g., precision farming, information systems, Internet, biotechnology) and coordinated governance structures (e.g., contracts, vertical integration). Efficient structures will be put in place to minimize the transaction costs of transferring new knowledge, thereby creating value to be shared by the source and the recipients.

As the private sector continues to expand its role in the transfer of new technical knowledge to the farming sector, LGUs will continue to distance themselves from their traditional clientele. LGUs cannot and should not compete with the private sector in such technology transfer activities. At the same time, opportunities for knowledge transfer to less traditional clients such as agribusiness decision makers, intermediaries and consultants, integrators, government bureaucrats, and regulators will likely expand for LGUs. The informational needs for such sophisticated clients will be more complex. They will likely involve an increasingly diverse set of disciplines, thus, expanding the opportunities for LGUs to integrate "cutting-edge" scientific knowledge from multiple disciplines into advanced knowledge packages (e.g. executive and continuing education packages, expert panels, think tanks).

Advanced knowledge generation and brokerage for selected decision-makers will likely define the new role of LGUs. Such knowledge will minimize transaction costs in agricultural technology transfer. The transaction costs incurred by the recipients will be minimized by means of knowledge packages. As well, the transaction costs incurred by LGUs will be minimized by means of coordinating with a smaller and more homogeneous set of recipients whose knowledge and technical needs will be better understood.

LGUs are well endowed with human capital to act as brokers of advanced scientific knowledge from diverse disciplines. The true challenge will be to create the structures that effectively integrate scientific knowledge from different disciplines while maintaining a strong disciplinary focus in knowledge-generating activities. Integrating knowledge from diverse disciplines is, of course, no trivial task. Strong asymmetries in knowledge exist among scientists due to differences in disciplinary backgrounds, cultures, and paradigms. Efficiency in knowledge exchange requires development of a common vocabulary to assist communication. Large stocks of knowledge on diverse problems, methods, and solutions must be exchanged to create a common knowledge base. The conditions of transfer are also important. Lack of understanding, trust, and mutual appreciation impose constraints in knowledge exchange and integration. Transaction costs associated with the integration of advanced knowledge from multiple disciplines will likely be significant. Thus, appropriate incentives and structures will have to be put in place to facilitate the integration of new knowledge into advanced knowledge packages.

APPENDIX A

MEASURING TRANSACTION COSTS IN LAND GRANT UNIVERSITY ACTIVITIES: A CASE STUDY

In this section we employ a case study to evaluate the proposition that transaction costs incurred by LGUs in knowledge transfer are large. We use University of Missouri-Columbia (UMC), a LGU, as our case. Through a mail survey sent to all scientists in the College of Agriculture of UMC, we collect information about the various knowledge-generating and transfer activities in which the scientists engage. One hundred and forty two agricultural scientists responded to our survey. This number represents a 30 percent response rate.

Using the survey information we measure transaction costs as the time of the agricultural scientist spent transacting for knowledge access and transfer. Clearly, additional transaction costs are incurred by the university in the process of technology generation and transfer. However, transaction costs related to the scientists' time will likely be the largest cost. In addition, as they are incurred directly by the scientists, transaction costs are powerful constraints.

We measure transaction costs following the approach of Wallis and North (1986). In particular, we first identify all transacting activities based on the behavioral model of technology transfer presented in section three. We subsequently add up time spent in transacting. Transaction costs are calculated as percentage of a scientist's total time on the job. Calculated transaction costs are reported below for the most important categories of activities in which agricultural scientists are typically involved.

Transaction Costs in knowledge access and transfer by activity

Activity	Research	Teaching	Extension	Service	Administration	Total
Transaction costs by activity (% time)	25.85	9.09	4.45	6.55	8.53	54.47

On average, a little over half (54.47 percent) of the total scientist's time on the job is devoted to transacting. This result is consistent with our basic proposition that transaction costs are large, both in an absolute and relative sense, and as such, they will affect the research strategies of agricultural scientists.

APPENDIX B

THE IMPACT OF TRANSACTION COSTS ON THE BEHAVIOR OF AGRICULTURAL SCIENTISTS

We evaluate whether agricultural scientists in LGUs have been increasingly focusing on disciplinary knowledge-adding activities emphasizing measurable academic outputs and paying less attention to contact with specific clientele groups. These behaviors are suggested by our behavioral model as researchers attempt to minimize transaction costs and respond to a changing incentive structure.

Emphasis of agricultural scientists towards academic research outputs and their limited contact with clientele groups was documented by Busch and Lacey (1983) over a decade ago. Since then, ample anecdotal evidence exists that such tendencies have continued, although additional formal empirical evidence has not been provided. UMC agricultural scientists perceive grants and contracts as well as publication in top academic journals to be the most important and rewarded activities in the current university environment (Table 1). Achieving highly-rated

Table 1. Perceived University Reward System

Activity	Perceived importance of activity[a]
Publishing in top journals	4.08
Publishing many journal articles	3.33
Seeking grants and contracts	4.54
Achieving high teaching evaluations	2.66
Providing significant service contribution	2.52
Establishing effective clientele relationships	3.06
Providing significant extension contribution	2.02

[a] Mean Scores are based upon 5 point scale (5=very important, 1=not important). Mean scores are estimated through regression analysis accounting for expected divergence of opinion due to differences in academic appointments among the respondents. Teaching, research, and extension appointments were explanatory variables of value perceptions for the various activities ranked.

teaching evaluations and significantly contributing to extension programs are considered less important.

Other scientists from their own discipline is the group that UMC agricultural scientists interact with the most (Table 2). Contact with client groups (e.g., farmers, agricultural businesses, non-profit organizations) and extension personnel is rather limited. Such results are consistent with those reported by Busch and Lacey (1983). The degree of contact is parallel to the influence that such individuals and groups have on the research agenda of UMC agricultural scientists. Strong focus in disciplinary research is further evident as a large part of their research agenda

Table 2. Contact with clientele and influence on research agenda

Contact persons/group	Degree of contact [a]	Degree of influence on research [b]
Scientists, own discipline	3.68	42.80
Scientists, other disciplines	2.96	16.65
Granting agencies	2.10	15.80
Private farm businesses	1.90	6.02
Extension personnel	2.21	5.04
Government officials	2.35	4.98
Farmers	2.37	4.91
Citizens and nonprofit organizations	1.93	3.54

[a] Mean score based upon 4 point scale (4=regularly, 1= hardly ever).
[b] Expressed as % of total research agenda.

is influenced by other scientists of the same discipline. All client groups taken together influence approximately 20 percent of their total research agenda.

We also evaluate our proposition that agricultural scientists tend to minimize contact with extension and client groups as they attempt to minimize transaction costs and maximize their productivity in disciplinary research that is most rewarded. If such argument holds, then those who have less contact with extension agents and clients should, all else equal, be more productive in disciplinary research. The respondents of the survey were asked to evaluate their productivity in a variety of academic outputs relative to their disciplinary standards. The level of contact with extension agents and client groups and its impact on disciplinary research productivity was analyzed through regression analysis. The estimated parameters along with their t-statistics are reported below:

$$\underset{(6.6)}{PUBL= 1.95} + \underset{(1.99)}{0.004*RA}+\underset{(2.31)}{0.11*RNK}+\underset{(2.32)}{0.22*DSC}+\underset{(0.89)}{0.07*OD} \underset{(1.89)}{-0.14*EX}+\underset{(0.51)}{0.04*FR}+\underset{(0.53)}{0.30*GR} \underset{(1.81)}{-0.14*ORG}$$

The results of the regression model indicate that publication levels in top disciplinary journals (PUBL) increase with the level of disciplinary contact (DSC) and decrease with the level of contact with extension personnel (EX) and non-profit organizations (ORG). The level of contact with scientists from other disciplines (OD), farmers (FR), and granting agencies (GR), do not have statistically significant effects on disciplinary research productivity. Publication productivity also increases with professional rank (RNK) and the research appointment of the scientist (RA). These results are in agreement with our proposition on interaction and coordination among scientists and client groups as well as the implied effects of associated transaction costs on disciplinary performance.

NOTES

[1] In any technology, transfer an individual or an organization may assume the role of both a source and a recipient in sequential knowledge exchanges. For example, an extension agent is a recipient when accessing new knowledge from an agricultural scientist on new varieties. By synthesizing such knowledge with his own knowledge of local growing conditions and transferring the processed information to farmers, the extension agent in turn becomes the source of new knowledge.

[2] Other factors may also contribute to increased transaction costs in agricultural technology transfer. For instance, property rights for new technical knowledge are loosely defined. Transaction costs associated with securing, monitoring, and enforcing such rights are often sizable. Examples of such costs include invention evaluation expenditures, patent application and maintenance fees, and monitoring and infringement litigation expenditures. The relative size of such transaction costs, however, is expected to be rather limited.

[3] This differentiation may be valid only when comparing value of information and knowledge at a given point in time. The tradeoffs between static efficiencies of information and dynamic cumulative effects and positive externalities of general knowledge are not delineated in this value differentiation.

REFERENCES

Abrahamson, E. 1991. Managerial fads, and fashions: The diffusion and rejection of innovations. Academy of Management Review 16:586-612.

Boehlje, M. 1994. Information: What is the public role? Staff Paper 94-17, Purdue University, Indiana.

Busch, L., and W. Lacey. 1983. Science, Agriculture and Politics of Research. Westview Press, Boulder Colorado.

Feder, G., R. Just, and D. Zilberman. 1985. Adoption of agricultural innovations in developing countries: A survey. Economic Development and Cultural Change, 254-97.

Huffman, W., and R. Just. 1995. Transactions costs, fads, and politically motivated misdirection in agricultural research. Working Paper, Iowa State University.

Huffman. W, and J. Miranowski. 1981. An economic analysis of expenditures on agricultural experiment station research. American Journal of Agricultural Economics, 63: 104-18.

Kline, S. 1991. Models of innovation and their policy consequences. In H. Inose et al. (eds.).

Ortmann, G., F. Patrick, W. Musser, and D. Doster. 1993. Use of private consultants and other sources of information by large cornbelt farmers. Agribusiness An International Journal, 9:391-402.

Pavitt, K. 1991. Technology transfer among the industrially advanced countries: An overview. International Technology Transfer 3-23.

Polanyi, M. 1966. The Tacit Dimension. London: Routledge and Kegan.

Rhodes, V.J. 1993. The industrialization of agriculture: Discussion. American Journal of Agricultural Economics 75:1137-9.

Rogers, E. 1983. Diffusion and Innovation, (2nd edition), New York, Free Press.

Rogers E., and F. Schoemaker. 1971. Communication of Innovations: A Cross Cultural Approach. New York: Free Press.

Rose-Ackerman, S., and R. Evenson. 1985. The political economy of agricultural research and extension: Grants, votes, and reappointment. American Journal of Agricultural Economics, 67:1-14.

Senker, J. 1995. Tacit knowledge and models of innovation. Industrial and Corporate Change, 4:425-447.

Stein, T. 1995. Background paper for Workshop on Privatization of Technology and Information Transfer in U.S. Agriculture, University of Wisconsin.

Wallis, J., and D. North. 1986. Measuring the transaction sector in the American economy 1870-1970. In Long Term Factors in American Economic Growth, S.L. Eugerman and R. Callman (eds.). Chicago University Press.

Williamson, O. 1989. Transaction costs economics. In Handbook of Industrial Organization, Schmalensee and R. Willing (eds.).Elsevier Science Publishers.

Wolf, S. 1995. Cropping systems, and conservation policy: The roles of agrichemical dealers and independent crop consultants. Journal of Soil and Water Conservation, 50:263-69.

14

The Changing Nature of Agricultural Markets: Implications for Privatization of Technology, Information Transfer, and Land Grant Research and Extension

David Zilberman, David Sunding, and Madhu Khanna

Public provision of research and extension has been the major feature of U.S. agriculture in the 20th century, helping maintain agriculture's competitive structure. However, new developments challenge the land grant system that provides public research and extension. These developments include changes in the nature of agricultural products and production firms, changes in technology, and a growing awareness of the importance of environmental considerations. These changes are most apparent in California, a state where agriculture is input intensive and which has an advanced technological infrastructure. These changes are also apparent in livestock industries, particularly in the poultry and swine industries. This paper analyzes these emerging phenomena, uses economic principles to interpret them, and assesses their impact on agricultural research and on the future of extension.

The paper first provides an overview of the current public research and extension system in the United States, an overview that is based on the literature but also reflects the authors' perspectives based on their experience in California. The paper then addresses six basic changes in American agriculture and discusses

ISBN 1-57444-104-3/98/$0.00/$.50

ways the public research and extension system can cope with them: (1) transition from production of commodities to production of differentiated products; (2) new types of farming and expanded range of farms; (3) increased reliance on specialized subcontractors and consultants for input use and management; (4) heightened concern with environmental quality and the side effects of agriculture; (5) emergence of institutions to transfer technology from the university to the private sector; and (6) the emergence of modern biotechnologies. The following six sections describe the basic issues raised by each type of change, and the concluding section discusses ways the land grant system can respond to these changes.

PUBLIC RESEARCH AND EXTENSION IN AGRICULTURE

Over the last 130 years, government has played a systematic and significant role in providing U.S. agriculture with research and extension services. Public research in agriculture has been conducted by U.S. Department of Agriculture researchers working in organizations such as the Agricultural Research Service and the Economic Research Service and through research activities at land grant universities. Faculty at the land grant universities spend part of their time as professors in specialized departments aimed to produce highly trained individuals specializing in agricultural sciences. Most faculty time is allocated to research in agricultural disciplines (agricultural economics, plant pathology, entomology, soil science, etc.). Research efforts address both basic and applied problems. There are land grant colleges in each of the fifty states, and these colleges are major sources of agricultural and environmental knowledge.

Two outstanding recent studies by Evanson and Huffman (1993) and by Alston and Pardey (1996) demonstrate quantitatively that the rate of return for agricultural research has been very high, and that this research has contributed significantly to the economic well-being of both U.S. and global agriculture. Alston and Pardey (1996) note a tendency towards reduction of public support for research in agriculture and, to some extent, substitution of private funding for public funding. They notice, however, that while effective transfer of knowledge to the private sector can unleash entrepreneurship and increase productivity, the private sector tends not to emphasize development of basic knowledge. Alston and Pardey warn, therefore, that continuing reduction in support for public research in agriculture will erode the accumulation of basic knowledge which is the key to development of future technology, and it will ultimately reduce the capacity of U.S. agriculture to address changes in environmental and economic conditions.

The organization of extension varies across states, and this paper looks at the future of extension using a model based on California that may apply to other states. There are two types of individuals employed by Cooperative Extension. Some work at the university as extension specialists and others are county agents. County agents are located throughout the state. Traditionally, educating farmers has been their major mission. At least in the past, much of their program has involved modifying production systems for individual farmers and the introduction of new technology through direct contact with growers. However, county agents also have played an important role in promoting collective action among farmers, recognizing opportunities at the regional level, initiating new directions for resource management and agricultural production, etc. The specialists provide a link between county agents and university basic researchers, and they have their own applied research program. They may deal with cross-county problems, develop statewide initiatives, work with federal agencies on state problems, etc.

Feder and Evanson (1991) overview the literature on the economics of extension, but to date there is not a comprehensive econometric study of the overall productivity of extension. It is clear that extension in the United States has made a significant contribution in diffusing technology. Huffman and Miranowski (1981) argue that the contribution of extension to the adoption of modern technology has been especially significant in regions where farmers lack relative formal education and extension activities can provide a good substitute for formal education. Furthermore, extension personnel have introduced new technologies and new crops to certain regions. For example, drip irrigation was introduced to California by two farm advisors, Gustafson and Hall (Caswell et al. 1984).

Extension personnel have also been important in promoting collective action and initiating institutions such as farmers' cooperatives, the Farm Bureau, and many other farmers' organizations. Furthermore, extension personnel have the knowledge and connections with research centers to be able to develop solutions to unexpected problems. For example, in some regions extension personnel have been important in developing solutions to environmental side effects of agricultural production. The dairy industry in Southern California was able to address concerns about water quality regulations due to technology modification and regulatory compromises that were enacted to a large extent through the efforts of an extension specialist, Jules Meyers, and other farm advisors.[1]

The outcomes of both research and extension activities are random and one needs a broad perspective to assess institutional performance. It is clear that relatively few projects yield most of the returns. In business school the rule of thumb is that 20 percent of activities generate 80 percent of profits, and this underestimates what is going on in research and education. At the University of California, ten innovations produced 90 percent of technology transfer revenues. Studying the performance of BARD (Binational American-Israel Research and Development Fund), Just et al. (1988) find that out of about 550 projects, approximately

50 have some commercial potential, and about ten produce most of the economic benefits. Thus, when redesigning public research and extension, one must identify the key components of success stories and maintain the more promising operations while modifying the rest. Furthermore, one must realize that a lot of the investment in such activities may not result in much, but a small percentage of successful ventures may lead to most of the returns.

THE TRANSITION FROM PRODUCTION OF COMMODITIES TO DIFFERENTIATED PRODUCTS

There are fundamental changes taking place in the definition of agricultural goods, and it is an open question how the land grant research and extension system will respond to, or even encourage, these developments. In particular, food products are increasingly differentiated. For example, consumers no longer simply purchase peaches. Instead shoppers select from an array of fresh peach products that are conventionally or organically grown direct from family farmers, pesticide-free, winter- and summer-season, and so on. In such a differentiated market, information plays a key role in helping consumers understand alternative product choices and in communicating these preferences to growers. In the future, there is likely to be growing demand for basic biological and technical knowledge to inform production processes that can yield well-defined products. As markets become unregulated and products more differentiated, there are growing opportunities for noncompetitive behavior, and the performance of such markets requires study and assessment. Provision of this information, knowledge, and assessment may provide new opportunities to the land grant system.

Most foodstuffs are homogeneous commodities traded in markets that are active and have many participants. Over the years, the average size of the most efficient farms producing agricultural commodities has increased as a result of increasing returns to scale associated with improved machinery and other technological innovations. Still, the average size of efficient units producing commodities is small with relation to the scale of the market. Therefore, competitive industries tend to conduct production of most agricultural commodities.

Knowledge and information are the most significant sources of returns to scale in agriculture. Both have what Romer (1986) calls "non-rivalry in consumption." Namely, they may be used by many individuals without being destroyed; therefore, it is optimal not to restrict the number of users of information. The cost of investing in knowledge and information is significant. Capturing investment in knowledge and information may require a much larger scale of operation than is needed to capture investment in capital goods such as tractors and other machinery. By providing public research and extension through the land grant system and other public institutions, the U.S. government eliminated the main source of scale in agriculture, enabling the competitive structure of American farming.[2] In fact, production of major agricultural commodities is quite competitive throughout the world. However, while the number of producers of major commodities is large, only a few producers have significant power in commodity markets. Concentration occurs primarily during the wholesaling and marketing of agricultural commodities.

Lancaster (1966) suggests that consumers do not derive enjoyment directly from the consumption of commodities that are in the market. In some cases, however, a production process occurs within the household that generates what he calls quality characteristics that provide consumer enjoyment. He suggests that product characteristics have intrinsic value and that goods sold in markets are, in essence, bundles of characteristics. The higher the value of the characteristics embodied in a good, the higher the value of the good. Thus, agricultural commodities are inputs in a production process that either takes place in the household or, as in the case of most industrial societies, in agribusiness firms. The final consumer-purchased products are far removed from the commodities produced in the field.

While there are few significant returns to scale in farming, there are significant increasing returns to scale in the transfer and marketing of agricultural commodities. This fact has led to situations where growers of food grains sell to a few buyers, who, in some cases, take advantage of their competitive situations. Furthermore, over time, processed food has become more elaborate, the value of final products has increased significantly, and the market shares of agricultural commodities in final food products have declined steadily. Today much of the value added in food production is generated beyond the farm gate.

Lancaster (1966), Rosen (1974), and other scholars who follow this line of reasoning recognize that per capita consumption of fruits and vegetables does not vary significantly with income; instead, product quality differs with income. High quality products are luxury goods, and their prices increase more than proportionally as quality increases. They also suggest that as income rises, people are willing to pay higher prices to avoid kitchen work such as cooking and cleaning. Their analysis suggests that agribusiness should follow a strategy of producing differentiated products with high variations in quality and degrees of labor saving and processing.

When consumers encounter a wide array of products with different qualities, they may have difficulty sorting them out and distinguishing between them. The introduction of brand names partially alleviates these informational and assessment difficulties (Carlton and Perloff 1990).

Brands allow easier identification of products and reduce consumer uncertainty regarding quality. With numerous brands, each owner receives some monopolistic power. Producers' monopolistic power increases with unique and/or essential products. Today final food products are sold in markets that can be described as monopolistically competitive (Carlton and Perloff 1990). While producers who operate in competitive markets are to a large extent price takers, producers who operate in monopolistically competitive markets can affect output prices. But the availability of close substitutes limits the extent of their monopoly power.

Thus far, the markets for major field crops have stayed competitive.[3] Historically, the relative prices of agricultural commodities have tended to decline over time, and the government established policies (commodity programs) to support the income of producers of major commodities (Gardner 1988; Tweeten 1974).[4] It has long been recognized that the increase in the value added of farm products is crucial to the prosperity of the U.S. agricultural sector (Cochrane 1979; Tweeten 1974).

While major field crops are bulk commodities with relatively small variations in quality, fruits and vegetables have become highly differentiated and their production modes and marketing structures have changed accordingly. Chemical additives modify the qualities and characteristics of processed fruits and vegetables, while changes in agronomical practices and crop varieties determine most of the qualities and characteristics of fresh produce.

Parker et al. (1991) developed a quantitative model to demonstrate that the price of fresh peaches may vary by seven- or eight-fold, depending on time of sale and quality characteristics such as size and sugar content. They argue that growers, by selecting variety and location (which affect time of harvest), can significantly affect product characteristics. Indeed, the ability to identify a market niche for highly profitable, differentiated products is a key element characterizing the production of fruits and vegetables. It has led to the cultivation of out-of-season fruits and vegetables in California, Florida, Arizona, and in recent years, Central America, Chile, and the Caribbean.

Emphasis on quality and product differentiation in fruit and vegetable production has led to the evolution of distinct marketing channels and government policies. Recognizing the importance of quality, growers fought early for the establishment of government marketing orders to enforce quality standards and provide growers with market power (Tweeten 1974).[5] Agribusiness firms that market fruits and vegetables play a key role in the emphasis on quality. These businesses provide growers with packing and sorting facilities and are responsible for transport and all the transactions associated with selling produce. In recent years they have increasingly relied upon contractual agreements with both growers and wholesalers and retailers (frequently large food chains). These contracts secure markets for the growers and specify production guidelines and quality and timing conditions. Similarly, contracts with buyers assure supply of consistently high quality agricultural products to food wholesalers and retailers.

The wine industry has been an archetype of adding value through product differentiation, quality control, and branding. Recently, some of the major fruit grower cooperatives (Sunkist and Blue Diamond) and large exporting companies (United Brands) have come to understand the importance of branding vegetables and developing contractual relationships with both producers and retailers. Over the last several years, this phenomenon has spread throughout the produce sector and aggressive new produce marketing firms have emerged. A distinguishing feature of these companies is their relentless pursuit of varieties with differentiated characteristics. Once they discover such varieties, companies contract with farmers to produce them and then establish a market through contracts with retailers and wholesalers. These types of companies have established new products such as kiwi (Sun World), yellow and red bell peppers, varieties of melons, etc.

Emphasis on value added and product differentiation is not restricted to fruit and vegetable production. The cut flower industry concentrates production in locations with favorable weather and/or in greenhouses that allow production throughout the year, in particular for Valentine's Day and Christmas, when production conditions are unfavorable but the demand for flowers is high (Dinar and Zilberman 1995). Growers have introduced traditional flowers (roses, carnations) in an increasing variety of colors, and they have developed alternative marketing channels, such as selling through supermarkets, florists, and mail order, with the price and quality of products varying between and across marketing channels.

Much emphasis on product differentiation and value added occurs in livestock production, in particular in the production of poultry. Large agribusiness firms, such as Tyson's, Foster Farms, and Purdue, establish contractual arrangements with producers whereby the agribusiness firms provide the genetic material, production guidelines, and sometimes feed, and then process and market under their own brand names.

Product differentiation and value added in agriculture are obtained not only by producing new products or developing new output characteristics, but also by distinguishing products according to the way they are produced. Some religions provide specific guidelines for food production and processing, and the size of the market segments served by producers who adhere to such guidelines may be significant (for example, the Kosher market). The extent and use of chemicals in production provides another set of criteria for product differentiation. Pesticide-free foods, free-range chicken, and beef raised without added hormones are example of differentiated specialties.

NEW TYPES OF AGRICULTURAL PRODUCTS AND EXPANDED RANGE OF FARMS

Along with product differentiation, in the future we will likely see significant expansion of activities that can be regarded as "agriculture." Agriculture can be viewed as consisting of land-intensive biological production systems, and agricultural activities can be defined as "growing, harvesting, or collecting the byproducts of living organisms in a land-intensive production system." According to this definition, a vineyard producing grapes in a land-intensive manner qualifies as a farm while a winery fermenting bacteria to generate wine does not (it is not land intensive). Recent developments in biological technology and the steady accumulation of biological knowledge is likely to lead to new types of farming. While to some people the new types of farming described below may sound far out and futuristic, the scientific knowledge to support their establishment already exists to a large extent.[6] Furthermore, there are several profitable firms that currently practice these types of farming. New types of farming will expand the range of products produced by agriculture and the range of organisms and species used in agricultural production.

Fine chemicals constitute a major category of product produced by new types of farming. These are high value products used in small quantities in a wide variety of processes, such as nutrients, food colorings, and ingredients for pharmaceuticals and diagnostic products.

Additional future farm product categories include exotic animals and plants, recreation, and bio-remediation. Table 1 provides a list of emerging farm activities.

Most new types of farming activities, especially the high tech ones, are likely to be the result of discoveries made by publicly supported research efforts in universities and other research institutions. Biologists and other scientists who make the discoveries on which new farming technologies are based try to scale up their experiments as much as they can, obtain patents for their ideas, and look for investors who will buy the rights to development. In many cases, private groups buy rights to commercialize a new concept that can lead to new forms of farming, but sometimes public money is behind development.

These new types of farming activities will increase the value of the natural resources used in agriculture, and will expand the range of businesses involved in agriculture. Some chemical and pharmaceutical companies are already actively

TABLE 1. New Types of Agricultural Activities

Activity	Definition	Examples
Algae culture	Cultivation of macro algae (seaweed) and micro algae to produce fine chemicals and food products.	Beta Carotene is derived from micro algae in Australia, Israel, and the U.S., with annual sales about $100 million (U.S.). There are experiments underway to use algae to produce food coloring and fatty acids.
"Pharming"	Producing pharmaceuticals through genetic manipulation of plants and domestic animals.	There are plans to raise animals to provide organs and to produce materials such as insulin. Furthermore, animals are fed to produce milk and eggs with unique medicinal properties.
Farming of exotic species	Domestication and cultivation of species of plants and animals that are now harvested (sometimes illegally) in the wild.	Cultivation of rhinoceroses is being considered by developers so the demand for materials produced from these animals will be met without destroying them in the wild. Similarly, Meyers suggests cultivation of exotic trees to slow destruction of forests.
Waste water remediation	Use of biological agents (algae plants) to purify waste water.	Rice and other plants are used to remove selenium from contaminated water. Oswold developed methods to use algae for sewage treatment.
Wetlands	Generation of artificial wetlands (wetlands produce vegetation for a variety of species in the wild).	Entrepreneurs are looking to obtain wetland development rights by buying ownership in new wetlands.
Fish farming	Cultivation of fish and other seafood in ponds and other bodies of water.	Fish and seafood farming generates billions of dollars globally (Thailand, Taiwan, and Ecuador are major producers). Significant fish farms operate in the South. U.S. fishery production has been enhanced by hatcheries.

involved in agriculture. These companies support ventures that develop new crops, taking responsibility for the production of new agricultural products. This science-based agriculture may require highly trained individuals, and in the future its exact structure will depend on training, research, and educational support provided by the state. Will this agricultural sector consist of many small farm units producing to meet the specifications of marketers or processors? Or will we see vertically integrated agribusinesses? The structure will depend to a large extent on environmental policies and waste management regulations, tax incentives for various types of organizations, crop insurance policies, etc.

Some of the new types of farming mentioned above may be appropriate at only a very few locations with particular characteristics, while other new types of farming may be applicable in a wide range of locations. The size and the structure of industries spawned by new farming activities depend on the demand for products and how much can be produced per acre. For example, world demand for certain fine chemicals may be easily supplied by one lab or by production on one or two acres of land, even though sales may generate a significant amount of money. Consequently, there will be one or two specialized manufacturers at most.

In the case of fatty acids, products are consumed most heavily by people who suffer from health problems or who want to protect themselves from such problems. The volume of the potential market may be in the billions of dollars, and production to meet world demand may require tens of thousands or even hundreds of thousands of acres. In this case, the industry is likely to consist of a relatively large number of producers with commercial characteristics, similar to those found in traditional commodity-producing agriculture. In the case of bison farming, signs are that production is spreading throughout the United States and it is likely to be quite competitive.

These new types of farming may provide a large array of opportunities for rural development. But development of these technologies is risky, likely requiring capital unavailable from the private sector. Thus, public support for research may be needed to enable development. Alston and Pardey (1996) and de Gorter and Zilberman (1990) argue that benefits from the introduction of new technologies are shared by the private entrepreneurs who earn profits through product sales and the public who benefits from lower prices, increased choices, and higher wages. Thus, in many cases, private sector research and development are suboptimal and government support for research and development is needed. The introduction of new types of farming will require applied research to adapt the technology to particular regional characteristics, and the public sector may need to support or supply some of this research.

Farm product differentiation and the development of new types of farming may expand the range of farm operations in the future. Our analysis and recent developments suggest that four major farm categories will evolve:

- *Commercial commodity-producing farms.* Most of the economically viable grain farms in the midwest fall into this category. We expect that these will continue to operate as family farms that control several thousands of acres and produce major food staples for world-wide commodity markets. The total number of these operations is likely to decline over time as individual units become larger, but overall, the structure of the grain industry will continue to be competitive in the foreseeable future. At the same time, we can expect some differentiation in product quality in major commodities and we will see the introduction of quality differentiated prices, but this will not alter the competitive structure.

- *Farming enterprises producing differentiated products.* This sector, discussed earlier, is likely to grow significantly and play a dominant role in fruit, vegetable, and some livestock operations. In this sector we will see an increase in contracting and vertical integration as well as monopolistically competitive differentiated markets.

- *New farming sector.* This category is comprised of the new types of farming activities discussed above. As we argued, the size of this industry, and even its market channels, may vary according to the type of product produced.

- *Small recreational farms.* Throughout the United States, there are more than a million operations that can be classified as small farms. These operations have annual revenues below $100,000 and, in many cases, operate on a part-time basis. They are part of a sector that we expect to survive and even expand. In particular, we expect to see a lot of such farms located on the urban fringe and in areas of natural beauty. In Europe, small farms in pastoral settings are sometimes subsidized. These operations provide aesthetic benefits in addition to their production benefits. These small operations also benefit their owners by providing a desirable lifestyle.[7] Furthermore, in many places, small farms are part of a thriving tourism industry. For example, Napa County in California is a region that combines small farming with tourist activities.

Distinction between the four types of farm categories may, in reality, be blurred. Many small farms, especially those producing fruits and vegetables, may produce differentiated products according to contractual agreements with larger farms or packers and processors. Many small farms may also be involved in new types of farming activities. Some of the large commodity-producing farms will also diversify their land to produce differentiated products. Nevertheless, these categories provide a useful typology in distinguishing the impact of policies on different elements of the farm sector and assessing the future of research and education in agriculture.

INCREASED USE OF SPECIALIZED SUBCONTRACTORS AND CONSULTANTS FOR INPUT USE AND MANAGEMENT

Increases in specialized knowledge associated with agricultural input use, the high cost of specialized inputs, and the complexity of legislation affecting agriculture has led to increased reliance on specialists to subcontract agricultural activities and to provide advice and guidance. Some of these tendencies are manifested already in high-value, input-intensive crops in California and Florida, and they are spreading throughout the country. These tendencies are likely to increase in the future and become characteristic of farming in many regions.

In the past, the high cost of machinery frequently led to the use of contract harvesting in many parts of the country. Today, the high cost of land preparation and leveling equipment is leading farmers to use subcontractors for land preparation activities. In California, the complexity of labor management and safety regulations has led growers to rely increasingly on subcontractors for harvesting activities (Gabbard 1993).

Custom services were established to address scale problems in agricultural production. While larger farmers might own specialized inputs, for example, laser levelers or aerial spraying equipment, midsize and small farmers rent this equipment. In fact, farmers have used custom services for years for aerial spraying. As more high-cost specialized equipment is introduced, and as long as some midsize and small farms are viable, the range of capital-intensive custom services will increase.

In addition to contracting out activities that require specialized capital equipment, farmers have contracted out activities that require specialized human capital. An obvious example is reliance on veterinarians for animal management. But scientific and technical development have increased the range of knowledge-intensive activities that are contracted out. The use of breeding services is commonplace. There is growing reliance on consultants for the design of feed formulas. In addition to increased reliance on consultants to address specific aspects of farm management, there is growing reliance on management consultants (especially for high value crops) to provide advice on all aspects of production and marketing. In many cases, management companies actually take over managing farms, especially when farming operations are owned by financial institutions or individuals who are removed from farming.

In many cases, the use of specialized experts is associated with the introduction of new management technologies. The introduction of modern irrigation technologies (sprinkler, drip, centrifugal) that increase irrigation precision and allow the application of other inputs through the irrigation system, has led to increased reliance on expert advice in irrigation. The design of irrigation systems is quite challenging and has become a specialized activity (sometimes provided by dealers and sometimes by independent consultants). Furthermore, at least in California, we have seen the emergence of irrigation consultants who combine sophisticated software and detailed weather information to provide irrigation management advice. The range of services varies from installment and maintenance of computerized management systems to day-to-day involvement in water management activities.

As monitoring technologies improve, there is a growing range of management services provided by consultants (some are independent professionals and some are employees of chemical companies) using new soil-testing technologies (Babcock 1992) and advanced information technologies. These consultants are important in the introduction of precision-farming technologies that rely on satellite information to adjust input use by location. The use of professional consultants is especially prevalent for pesticide applications. In studies in California, Wiebers (1990) finds that in some situations (insect control) a plurality of farmers may use consultants. While the number of independent pesticide consultants in California is significant, more consultants are employed by pesticide dealers. Campbell finds that the reputation of consultants is a dominant factor in farmers' use of consultants and even pesticide dealers' choices. Furthermore, there is a growing tendency to pay consultants a fixed fee rather than a commission. Concern for reputation and a fixed salary are likely to reduce the incentive for consultants working for dealers to over prescribe pesticides. Consultants, employed by dealers of agricultural chemicals, play an important role in introducing precision technologies that are supposed to reduce chemical use and increase yields. The effort consultants invest in promoting precision technologies depends to a large extent on the impact of adoption on dealers' income. In particular, it depends on the profit margins from sales of chemicals versus leasing and sales of precision equipment and information.

The increased role and importance of private consultants make some of the public sector farmer education programs redundant. Efficiency arguments suggest that public activities that compete with private consultants should be reduced or eliminated and instead the public sector should develop a program that has public good properties that complement the activities of private consultants. Furthermore, the legal responsibilities of private consultants may need to be defined, especially in terms of liability (are consultants liable for bad advice or malpractice?). Private consultants may also provide a better means to address policy concerns, in particular in the environmental area, as we discuss below.

ENVIRONMENTAL QUALITY AND THE SIDE EFFECTS OF AGRICULTURE

While the tendency toward differentiated products leads to increasingly privatized agricultural systems, concerns over the environmental and human health side effects of agriculture suggest that unchecked private pursuit of profit may lead to overuse of agricultural inputs (Carlson et al. 1993). Agricultural production has led to a variety of environmental problems, including air pollution caused by burning rice and plowing, soil erosion, and contaminated water due to the runoff of nitrates, other fertilizers, and animal waste.

According to economics, the environmental side effects of agricultural production are "externality problems."[8] These side effects cannot be resolved by market forces alone and require market intervention through incentives such as taxation, direct regulation, or establishment and enactment of explicit liability rules and/or property rights (Carlson et al. 1993). Indeed, efforts to introduce policies to address the environmental side effects of agricultural production have intensified significantly in recent years.

There are several obstacles to establishing effective policies to address environmental problems in agriculture. First, there is the multidimensional nature of side effects and the heterogeneity of agriculture. The environmental side effects of agriculture are complex. For example, pesticide use results in several possible side effects, including food safety problems, worker safety problems, ground water contamination, and health risks to many species and organisms in the environment.

The magnitude of environmental side effects depends very much on location and the manner in which activities are conducted. In the case of pesticides, the impact of environmental side effects is determined by the amount and type of chemical used, the application process, and the characteristics of the environment. For example, aerial spraying of pesticides results in more severe environmental side effects than spraying with precision equipment. Application of pesticides on sandy soil may lead to more ground water contamination than application on heavier soil. Some pesticide problems (e.g., negative effects on biodiversity or depletion of the ozone layer) are global in nature. Other problems, such as farm worker safety and crop injury due to pesticide drift, are local in nature.

The multiplicity of side effects associated with agricultural inputs, the dependence of their impact on the use of other inputs, and the variation of their impact across different locations make uniform across-the-board regulations suboptimal. For example, despite theoretical appeal, an across-the-board pesticide tax is unlikely to lead to optimal use of any chemical because damages vary by location.[9] Efficient incentives and regulations to address agricultural externality problems may vary by location, product, and application technology.[10] To attain effective and efficient solutions, regulatory agencies need to emphasize flexibility, consider options on a case-by-case basis, and adjust policies as technology and knowledge change over time.

A second obstacle to establishing effective policies to address environmental problems in agriculture stems from uncertainty and lack of information. There is a significant lack of knowledge regarding the processes that generate the environmental and health side effects of agricultural production. Many agricultural externality problems stem from residues from applied inputs. Environmental damage from these residues depends on the outcomes of three processes (Bogen 1985): transport (for example, movement of pesticide residue from the location in which the pesticide was applied to locations of possible impact); exposure; and dose-response (the relationship between damage to an organism and the amount of toxic material to which it has been exposed). Modeling these processes is complex and empirical estimation of their outcomes requires experimentation, measurement, and adjustment to specific situations.

Political economic considerations represent a third obstacle to establishing environmental policies for agricultural production. Uncertainty about the environmental side effects of agricultural production and the complexity of designing policy responses to such problems provides a fertile background for political bickering. Furthermore, regulation of agricultural activities affects many parties, including farmers, consumers, input manufacturers, and environmentalists. Even within each group there may be sub-groups who are affected differently. For example, banning a pesticide may impose severe economic costs on people who use this chemical but may benefit non-users (Zilberman et al. 1994). Even when there is a relative consensus regarding the need to address an agricultural externality problem, different groups may disagree about the means of addressing such problems and use the political process to affect outcomes (Buchanan and Tullock 1975). In the past, the debate about environmental policies has been litigious and confrontational, but there is a growing tendency to emphasize collaborative solutions based on consensus and establishment of wide base coalitions. However, establishment of collaborative outcomes requires establishment of agreeable and common information sources and existence of organizational capacity to promote collaboration and build coalitions.

On the other hand, some recent technological developments make it easier to enact effective policies to address environmental problems in agriculture. Recent and upcoming precision-enhancing technology in agriculture will reduce the cost of addressing agricultural externalities. With most agricultural technologies, there is a significant gap between the quantity of inputs applied and the quantity actually used by the crops. In many cases, the difference between applied and utilized inputs is the source of negative environmental side effects. Precision-improving technologies reduce the gap between applied and utilized inputs, but increases in the technical efficiency of input use (technical efficiency is measured as the ratio of utilized inputs to applied inputs) require more expensive application equipment or extra labor.

Modern irrigation technologies (sprinkler, drip, LEPA) are examples of precision-enhancing technologies. In California, drip irrigation may increase irrigation efficiency from sixty to nintey-five percent, but it requires a significant increase in fixed costs per acre. The use of scouts as part of integrated pest management is another example of enhanced precision technology. In contrast to preventive across-the-board applications, scouting helps to restrict pesticide applications to infected areas.

Recent improvements in communication and monitoring technologies have enabled the development of precision farming technologies that rely on information obtained from satellites to determine exact chemical application levels related to changes in field conditions. There is a growing literature on the economics and adoption of precision technologies.[11] Khanna and Zilberman (1996) argue analytically that profit-maximizing farmers who switch from traditional to precision-enhancing technologies are likely to have higher yields and, under plausible circumstances, reduce their input use and residue levels. They further argue that increases in output prices, input prices, and more strict regulations of residues are likely to increase adoption rates of precision technologies. Furthermore, the gains from switching to precision technologies vary across location, and the likelihood of their adoption is higher in regions in which the technical efficiency of existing technology is especially low.[12] Moreover, the specification and the design of the most effective precision technologies vary across location.

A second factor that may help in reducing environmental side effects of agricultural production is the biotechnological development of varieties that are disease resistant. Such varieties will reduce the need to use chemicals to address many problems. For many crops, a significant number of disease resistant varieties may need to be developed to address specific locational conditions. Furthermore, the introduction of such varieties may require adjustments in several other facets of the production process beside pest control.

A third factor that is likely to assist in facilitating policies to reduce agricultural externalities is the increasing reliance on specialized consultants mentioned above. One approach to allay problems associated with misuse of pesticides is to limit the right to prescribe certain chemicals (the more toxic ones) to certified consultants only, and to require reporting such applications. To build alliances, consultants will have to go through a rigorous educational program so they are best able to consider plausible alternatives and the economic and environmental impacts of the proposed pesticide application. California takes this approach with some of the more toxic chemicals. Other states are considering the idea of "plant doctors" to be part of an overall strategy to address pesticide control and pesticide problems.

The complexity of the relationship between agricultural activities and environmental quality and the high degree of uncertainty about these relationships suggests that local level research may be needed in order to establish environmental policies to control agricultural externalities. Research should address specific problems at the locations where environmental problems occur. In addition, debate regarding the establishment of policies to regulate environmental activities requires educating and negotiating with a multiplicity of interest groups. Thus, the policy process requires some institutional capacity to conduct applied research on local agricultural and environmental conditions, educate, and provide mediation and negotiation services.

THE EMERGENCE OF INSTITUTIONS TO TRANSFER TECHNOLOGY FROM THE UNIVERSITY TO THE PRIVATE SECTOR

Cochrane (1979) argues that rapid technological change is a major characteristic of modern U.S. agriculture. As Ruttan and others argue, new technologies are induced by economic realities and institutional set-ups and the pace of technological change is strongly dependent on policy and institutions.

Until 50 years ago, most agricultural innovations came from practitioners. For example, a farmer, John Deere, invented the iron plow which led to the industrial empire that contributed many mechanical innovations in agriculture (Cochrane 1979). Today, however, many chemical, biological, and informational innovations are science-based and have resulted from public-sector research. The basic concepts behind some key chemical innovations were discovered by university scientists. Chemical companies subsequently developed and commercialized these technologies. Over time, chemical companies have developed their own research labs and these labs have developed many new types of pesticides. Similarly, USDA, ARS, and the land grant system developed many basic seed varieties but, over the years, commercial seed breeders have developed their own varieties, relying in many cases on the material developed by public-sector research as their starting point.

Thus, over the years there has been continuous transfer of knowledge from the public to the private sector and coexistence of public and private research institutes. Yet, in general, public research institutes address more basic and fundamental problems and they have a broader horizon than most private research facilities.

A similar pattern of technology transfer and coexistence of private and public research has existed outside agriculture. However, over the last 20 years, concern about support for public research and education and issues related to equity and the distribution of gains from university-developed innovations has led to the establishment of offices of technology transfer at many universities. These offices formalize the process of technology transfer from the university to the private sector, and they have played a major role in transferring technologies in communications, electronics, and chemistry. But their most important impact has been in medical biotechnology.

Interviewing professionals in several offices of technology transfer, Postlewait et al. (1993) find that the individuals who run these offices do not use technology transfer as a mechanism to enhance university profitability per se. Rather, their main objective is to increase the use of university technologies in the private sector. Technology transfer officers recognized that private companies have not fully utilized university-developed technologies, and technology transfer officers are attempting to improve this situation. A major activity of the offices of technology transfer is thus to identify university discoveries that have commercial potential, issue patents for these discoveries, and then sell the rights to use these patents to private enterprises.

During 1994, U.S. universities earned about $350 million from various technology transfer activities. However, this sum is small in comparison to the more than $10 billion in university research expenditures nationwide. The University of California is the university with the highest earnings from technology transfer activities ($64 million in 1994, or about five percent of the university's research budget). Thus, even at the most successful research universities, technology transfer cannot cover the cost of public research.

Technology transfer revenues at universities are divided among the faculty (the discoverer), the university, and the individual department according to various formulas. Income from technology transfer agreements is extremely helpful to some departments, and can be used to upgrade their programs significantly. Some faculty members benefit significantly from royalties as well as from their income as partners and/or consultants to new biotechnology firms.

Technology transfer agreements are only one dimension of the relationship between universities and private firms. Private companies provide grants to support the university and the university educates future researchers and other personnel who work in industry. Indeed, we have seen the build-up of large industrial parks near research universities. Silicon Valley near Stanford is the most obvious example.

Technology transfer offices attempt aggressively to sell the rights to use patents to major corporations, but some of their outstanding successes involve transferring technologies and assisting in the establishment of upstart companies. The three largest medical biotechnology companies in the United States, Genentech, Amgene, and Chiron, are upstart companies that were founded by collaboration between venture capitalists and university professors. The annual revenues of these companies are now in the billions of dollars, and they are being taken over by giant pharmaceutical companies that have relative advantage in production and marketing of the medical products these biotech companies produce.

EMERGENCE OF BIOTECHNOLOGY

According to Huttner et al. (1995), biotechnology provides an advanced set of tools to use living organisms or parts of organisms (such as DNA or enzymes) to make or modify products. The most familiar applications are in medicine. These include human insulin, a clot-dissolving treatment, an effective new vaccine for hepatitis, and precise genetic diagnostic tools. The medical biotechnology industry has grown significantly over the last 20 years. In 1993, drugs produced by the industry achieved total sales of $7.7 billion. Medical biotechnology innovations comprise the leading source of revenues for university offices of technology transfer. For example, in 1994, at least five out of the 10 leading royalty-generating products for university offices of technology transfer were innovations related to medical applications, with human insulin leading the pack generating approximately $35 million dollars in royalties in 1994 (AUTM 1995; Huttner 1995).

Huttner et al.'s (1995) survey of the state of biotechnology argues that the bulk of the first wave of biotechnology innovations was medical and that agricultural innovations will play a major part in a second wave of innovations, likely to occur within the next twenty years. Biotechnology developments may critically change American agriculture. Biotechnology accelerates the process of plant breeding and allows modification of the characteristics of plants and livestock. Therefore, biotechnological developments will be crucial in the introduction of new differentiated products in agriculture. Researchers are already working on developing some varieties of fruits and vegetables that have a longer shelf life and enhanced flavor. Biotechnology will play a critical role in developing new types of farming activities and farm organisms and animals that have not been cultivated traditionally. As we mention above, biotechnology will also enhance pest control possibilities by developing pest resistant varieties and expanding the effectiveness of existing pest controls.

Although agricultural biotechnology is still in its infancy, there have been several successful applications. One example is the introduction of the *bacillus thuringiensis* (Bt), into over 50 plant crops to control pests. Ollinger and Pope (1995) find that most commercial applications of plant biotechnology, as well as most experimental trials that are close to commercialization, are based on innovations that enhance plant protection (for example, pest-resistant varieties or herbicide-resistant varieties). These applications are targeted at major crops and are undertaken by large pesticide, chemical, and seed companies. In many cases, the development of plant protection biotechnologies is not as expensive as the development of new crop varieties or products, and the chemical companies can use their marketing channels for promotion and sales. Furthermore, increased severity of chemical pesticide regulations threatens the economic viability of some major pesticide products, and manufacturers view biotechnology as an avenue for production of alternative products.

There have been fewer innovations that enhance final product quality (for example, flavor, color, or size), and, to a large extent, they have been undertaken by upstart companies. Biotechnology firms specializing in genetically engineered fruits and vegetables, such as Calgene and Dynep, have already introduced several new varieties. But, because of the high cost of developing new varieties and the difficulty in capturing the investment by sale of seeds, these companies have decided to develop their own brands of produce which will be sold in retail outlets. They aim to capture the economic surplus from the sale of their improved varieties and subcontract farmers to grow their products.

Some of the larger seed and chemical companies are likely to take advantage of new product-enhancing biotechnology innovations, but as of now, much of the research and development is left to upstart companies. As these companies succeed, they are bought out by established agribusiness conglomerates (that is what happened to Dynep and Calgene). The development of new biotechnology crops may also be supported by major produce marketing firms. Several of the large marketers of fruits and vegetables have shown an interest in buying the patent rights to genetically engineered varieties of fruits and vegetables, establishing brand names, and subsequently contracting for production through growers.[13] While these developments are promising, research and development of bioengineering for new crops is dwarfed relative to other areas of biotechnology.

There are several explanations for why agricultural biotechnology lags behind medical biotechnology. First, according to Huttner et al. (1995), the basic knowledge needed for agricultural biotechnology is less well understood than that needed for medical biotechnology. Medical biotechnology is based on knowledge regarding human health, which has been studied more intensively than the biology of the thousands of species of plants and animals that may be used in agricultural biotechnology. Second, medical biotechnology mostly emphasizes preventing certain symptoms, while some agricultural biotechnology innovations require significant modification of organisms. Third, medical innovations do not have to be adjusted to conditions at specific locations, while innovations of, say, new seeds, may require changes in production systems and adaptation to conditions in various locations.

Market and regulatory conditions also contributed to the earlier growth of medical biotechnology (Huttner et al. 1995). The order of magnitude of the market for new medical products appears to be more significant than the market for many agricultural innovations. The marketing of a new drug is relatively simpler than the marketing of a new agricultural innovation, especially of a new agricultural product. Finally, the regulation of agricultural biotechnology is restrictive and costly and significantly increases investment costs. It may be that there is greater social acceptance of genetic manipulation for medical purposes than for agricultural purposes, and this has resulted in a better regulatory climate for medical biotechnology. In sum, the greater knowledge base, larger market potential, and fewer complications in adaptations and marketing have given medical biotechnology a head start over agricultural biotechnology and have made it more attractive for investors and commercialization.

IMPLICATIONS FOR PUBLIC RESEARCH AND EXTENSION

Public research and extension have to adjust to the new changes affecting agriculture. The likely increase in product differentiation and emergence of monopolistically competitive agricultural markets will generate demand for exclusive rights to agricultural technologies and, in particular, crop varieties. Agribusinesses, which invest millions of dollars in establishing brand names for their agricultural products, are likely to have exclusive rights to some of the varieties they intend to develop and promote. Agricultural biotechnology tools are likely to provide new mechanisms to generate such appropriable technologies.

The evolving offices of technology transfer may play a crucial role in enabling privatization of the rights to agricultural technology developed by the University. The pace of development of agricultural industries with differentiated products depends critically on the rate of new innovation. Thus, continuation and even expansion of basic public research on plants, livestock, and agricultural systems will provide the foundation for new discoveries and inventions enabling development of differentiated agricultural products.

The availability of new knowledge will be crucial in setting the pace for establishing new types of agricultural activities. Basic research needed for such developments may emphasize the study of organisms not previously emphasized by scientists within the agricultural land grant colleges. Thus, to facilitate the development of new agricultural activities, agricultural research support should be extended to include basic research on organisms that have not been covered by the agricultural research infrastructure in the past. Hopefully, this research will lead to discoveries that, through technology transfer, will be appropriated and lead to development of new industries.

It was argued that the vigor and effectiveness of current technology transfer efforts in agriculture are hampered by the insufficient development of new innovations and the need to adapt new technologies to varying conditions. The scaling up of ideas to provide transferable innovations is likely to be another area where public research and, in particular, Extension should play an important role. Furthermore, there will be a growing need to provide information and facilities that will enable public sector investors to identify conditions and locations where new agricultural activities or new varieties of existing products can be produced profitably. This is another area where public efforts can provide significant returns.

The increased role played by private providers of agricultural inputs in instruction and education of farmers in the use of new technologies and the emergence of private consultants are likely to reduce significantly the role of the public sector in providing support to individual farmers for management of agricultural technologies. On the other hand, the efficiency and effectiveness of private consultants would be enhanced if the public sector would provide those consultants with certification, technology assessments, and improved channels of communication with researchers. Increased emphasis on the environmental side effects of agriculture and concern for environmental health will require significant increase of knowledge on the relationship between agriculture and the environment. Effective policies in this area will depend on expanded knowledge in basic biological, physical, economic, and social phenomena as well as knowledge regarding performance of the system in specific locations. Furthermore, there is and will be a growing need for education of farmers, government officials, environmentalists, and the general public on the relationship between agriculture and the environment and effective mechanisms to address problems in these areas. There will be growing demand for the establishment of institutions and mechanisms to facilitate collaboration in addressing environmental problems in agriculture. These are areas for further public-sector activities.

The changing nature of agricultural markets will require significant adjustments in the responsibilities and emphases of researchers in the Agricultural Experiment Station, Extension specialists, and farm advisors. For public research in agriculture, these changes include the following:

- *Continuation of the public basic research on biology and disease control of agricultural commodities.*

- *Expansion of the umbrella of the experiment stations to support the study of biology and properties of organisms that can lead to new types of farming.*

  Experiment station findings have to be available to scientists outside the traditional agricultural disciplines—biologists, botanists, and other scientists—whose research has the potential to develop new agricultural activities.

- *Expansion of the basic research on the generation and management of environmental side effects of agriculture.*

  This research will aim to: (a) Improve the understanding of the relationship of agricultural production and pollution (this includes a better understanding of processes such as pollution generation, contamination, transport, and exposure); (b) provide the foundation for improved monitoring and input-use technologies; (c) improve understanding and evaluation of choices in agricultural production and agricultural pollution control; and (d) evaluate alternative policies and institutions to provide the basic knowledge for the design of efficient and effective institutional policies.

- *Pursuit of agreements for obtaining partnership and cost-sharing from private sector agents benefiting from their research.*

  Changes in agricultural markets are likely to have significant effects on the performance of Cooperative Extension. It seems that the roles and responsibilities of Extension specialists located on campuses are likely to expand, while the role of the county agents may decline. Some of the activities of Extension specialists in the future should include:

- *Conducting research to expand knowledge; scaling up innovations obtained by basic research; and developing technology packages that will be transferable to the private sector for further commercialization.*

- *Initiating, managing, and conducting research projects to address agricultural and natural resource problems of their state.*

  Some of these projects may modify and adapt new discoveries to generate tailor-made innovations appropriate for specific weather conditions, crops, and production systems. Other projects may be policy-oriented and aim to provide policymakers with information on, and possible solutions for, agricultural and related environmental problems in the state. The specialists should engage other university researchers in their projects whenever appropriate and in some cases serve as initiators and managers.

- *Initiating educational programs for policymakers in private and public sectors to solve agricultural and environmental problems.*

  There should be a broad clientele for these activities including policymakers at the state and federal agencies and in agriculture and agribusiness, city planners, and environmentalists. They should also educate investors (foreign and domestic) about their state's agriculture, economy, and environment.

- *Initiating and overseeing employment training programs to prepare individuals for the new needs of agriculture.*

  Specialists should initiate interesting and unique training programs to enable farmers, policymakers, and professionals to adjust to the new technological realities. These programs should take advantage of new opportunities provided by the development of new information technologies.
  The changes in agricultural markets should drastically affect the roles of county agents and will likely reduce their ranks significantly. Adjustments in the roles of county agents should include:

- *Reducing, or even eliminating, instruction to individual farmers, especially when private sector consulting and education are available.*

  Extension agents should not compete with the consultants on advising the commercial farmers; rather, they should serve constituencies not served by the private consultants.[14]

- *Providing education to consultants and chemical dealers and providing objective assessments of new products and technologies.* County agents can serve as the middlemen for private consultants and university researchers. They can develop training and certification programs for activities that require location-specific knowledge. They can also advise consultants on the merits of alternative management strategies and products.[15]

- *Providing services in adapting technologies and production systems to local conditions and providing information and knowledge of local conditions to researchers and investment organizations.* The county agent should be a resource available to developers of new technologies (e.g., seed companies) which enable firms to experiment with new technologies in different environments and integrate them into existing production systems. These types of activities should be primarily financed by technology developers.

- *Mediating between groups and developing cooperative solutions to conflicts relating to resource use and implementation of environmental policy.*

NOTES

[1] Personal conversation with Jerry Siebert, ex-director of Cooperative Extension in California, and Bob McKuen, a leading farmer in the Santa Ana River basin in California.

[2] Cochrane argues that the competitive structure and the family farm emerged in regions of North America where the government provided the public good (the North prior to the Civil War), while regions without public good provision (the South, Mexico) witnessed the emergence of plantations and latifundia.

[3] Within each commodity, there are distinct markets reflecting variety and quality differences (long fiber cotton is distinct from short fiber cotton, Durham wheat from standard wheat). There is a growing tendency to pay feed producers according to nutritional content and quality. These quality considerations are accommodated within competitive markets, as long as the magnitude of variability is not sufficiently large (relative to the farming population) to lead to situations where a small number of producers establish monopolistic power in a captive market.

[4] Gardner (1988) argues that an increase in scale of farming has led average income at the farm sector to be on par or to exceed that in the nonfarm sector. Thus, the chronic budget deficit and international trade agreements (GATT, NAFTA) are main causes for a gradual phase out of major commodity programs.

[5] Marketing orders also serve as mechanisms to obtain collective market power for growers.

[6] Based on conversations with many biologists and consulting work, we draw heavily on what we learned from Dr. Suzanne Huttner, director of the Center of Biotechnology at the Univer-

sity of California; Dr. Shoshanna Arad at Ben Gurion University in Israel; Professors Michael Freeling, Bob Buchanon, Alexander Glazer, and George Poinar Jr. from U. C. Berkeley; and Professor John Buchikvwitzka from the University of Western Australia.

[7] Indeed, we have seen a significant expansion of the number of recreational ranchettes in California near the major cities and in Texas.

[8] Externality occurs when production activities result in unintended side effects on third parties.

[9] A uniform tax on pesticides may be optimal when the environmental side effects of pesticides are global. For example, it may be optimal to consider a uniform tax on methyl bromide residues if the main concern is the negative side effects of this chemical on the ozone layer (Yarkin, Sunding, Zilberman, and Siebert 1994). But even in the case of methyl bromide, there may be other reasons to develop differentiated taxes or other differentiated policies.

[10] Efficiency is defined here as maximization of net economic benefits taking into account externality considerations (Just, Hueth, and Schmitz 1982).

[11] See Boggess, Lacewell, and Zilberman's (1993) discussion on the economics of the adoption of modern irrigation technology.

[12] For example, drip irrigation is more likely to be adopted in locations with sandy soils or on steep hills where irrigation efficiency under gravitational technology is especially low. Indeed, this has been shown by a large number of studies (see Caswell 1991).

[13] Based on interviews with staff at the Office of Technology Transfer, University of California.

[14] One group of likely customers are small traditional farms, as long as there is agreement that such activities provide recreational and aesthetic externalities to the society. But, the issue of service fees is likely to be problematic.

[15] In some regions of California, where dealers and independent consultants provide much of the information to farmers, these dealers and consultants became an important clientele of county extension agents (Goldman, Shah, and Zilberman 1990).

REFERENCES

Alston, J., and P. Pardey. 1996. The Economics of Agricultural Research and Development. Washington, DC: AEI Press.

The Association of University Technology Managers, Inc. 1995. AUTUM Licensing Survey: FY 1994 Survey Summary and Selected Data FY 1991-1994. Norwalk, Connecticut.

Babcock, B.A., and A.M. Blackmer. 1992. The value of reducing temporal input nonuniformities. Journal of Agricultural and Resource Economics, 17:335-347.

Baumol, W. J., and W. E. Oates. 1974. The Theory of Environmental Policy. Englewood Cliffs, New Jersey: Prentice Hall.

Bogen, K. 1985. Uncertainty in environmental health risk assessment: A framework for analysis and application to a risk assessment involving chronic carcinogen exposure. Unpublished Ph.D. dissertation, University of California, Berkeley.

Boggess, W., R. Lacewell, and D. Zilberman. 1993. Economics of water use in agriculture. In Agricultural and Environmental Resource Economics, Gerald A. Carlson, David Zilberman, and John A. Miranowski (eds.). New York: Oxford University Press.

Buchanan, J. M., and G. Tullock. 1975. Polluters' profits and political response: Direct control versus taxes. American Economic Review, 65:139-147.

Campbell, M. 1993. The pesticide use network: An analysis of the relationships which influence pesticide use decisions. Berkeley, Management Systems Research.

Carlson, G. A., D. Zilberman, and J. A. Miranowski, editors. 1993. Agricultural and Environmental Resource Economics. New York: Oxford University Press.

Carlton, D. W., and J. M. Perloff. 1990. Modern Industrial Organization. Glenville, Illinois: Scott, Foresman/Little, Brown.

Caswell, M. F. 1991. Irrigation technology adoption decisions: Empirical evidence. In The Economics and Management of Water and Drainage in Agriculture, A. Dinar and D. Zilberman (eds.). Norwell, Massachusetts: Kluwer Academic Publishers.

Caswell, M., D. Zilberman, and G. E. Goldman. 1984. Economic implications of drip irrigation. California Agriculture, 38:4 and 5.

Cochrane, W. W. 1979. The Development of American Agriculture. Minneapolis: University of Minnesota Press.

de Gorter, H., and D. Zilberman. 1990. On the political economy of public good inputs in agriculture. American Journal of Agricultural Economics, 72:131-137.

Dinar, A., and D. Zilberman. 1995. The impact of crop protection chemical regulations on the cut flower industry in California. Gold River, California: Report prepared for the California Cut Flower Commission, Summer.

Feder, G., and R. E. Evenson. 1991. The economic impact of agricultural extension. Economic Development And Cultural Change 39:607-650.

Foley Scheuring, A. 1988. A sustaining comradeship: The history of University of California cooperative extension, 1913-1988. Division of Agriculture and Natural Resources. University of California, Berkeley.

Gabbard, S. M. 1993. Farm workers' labor supply decisions. Unpublished Ph.D. dissertation, University of California, Berkeley.

Gardner, B. L. 1988. The Economics of Agricultural Policies. New York: Macmillan Publishing Company.

Goldman, G., F. Shah, and D. Zilberman. A preliminary report on the economic analysis of University of California cooperative extension activities: The case of Stanislaus County, California. Department of Agricultural and Resource Economics, University of California, Berkeley.

Huffman, W. C., and R. E. Evenson. 1993. Science for Agriculture: A Long Term Perspective. Ames: Iowa State University Press.

Huffman, W. E., and J. A. Miranowski. 1981 An economic analysis of expenditures on agricultural experiment station research. American Journal of Agricultural Economics, 63:104-118.

Huttner, S. L., H. I. Miller, and P. L. Lemaux. 1995 U.S. Agricultural Biotechnology: Status and Prospect, Technological Forecasting and Social Change. Special Issue on Biotechnology and the Future of Agriculture and Natural Resources, D. Parker and D. Zilberman (eds.). 50:25-39.

Just, R. E., D. L. Hueth, and A. Schmitz. 1982. Applied Welfare Economics and Public Policy. Englewood Cliffs, New Jersey: Prentice Hall.

Just, R. E., D. Zilberman, D. Parker, and M. Phillips. 1988. The economic impacts of BARD research on the U.S. Report prepared for the Commission to Evaluate BARD.

Kerr, N. A. 1987. The legacy: A centennial history of the state agricultural experiment stations, 1887-1987. Columbia: Missouri Agricultural Experiment Station.

Khanna, M., and D. Zilberman. 1996. Technology adoption for environmental quality control. Unpublished Manuscript. Institute for Environmental Studies, University of Illinois.

Lancaster, K. J. 1966. A new approach to consumer theory. Journal of Political Economy, 74:132-157.

Meyers, N. 1992. The Primary Source: Tropical Forests and Our Future. New York: W. W. Norton.

Ollinger, M., and L. Pope. 1995. Strategic research interests, organizational behavior, and the emerging market for the products of plant biotechnology. Technological Forecasting and Social Change, 50:55-68.

Parker, D. D., D. Zilberman, and K. Moulton. 1991. How quality relates to price in California fresh peaches. California Agriculture, 45:14-16.

Postlewait, A., D. D. Parker, and D. Zilberman. 1993. The advent of biotechnology and technology transfer in agriculture. Technology Forecasting and Social Change, 43:271-287.

Romer, P. 1986. Increasing returns and long-run growth. Journal of Political Economy, 94:1002-1037.

Rosen, S. 1974. Hedonic prices and implicit markets: Product differentiation in perfect competition. Journal of Political Economy, 82:34-55.

Tweeten, L. 1974. Foundations of Farm Policy. Lincoln: University of Nebraska Press.

Wiebers, C. 1990. Economic and environmental effects of pest management information and pesticides: The case of processing tomatoes in California. Unpublished Ph.D. dissertation, Technischen Universitat, Berlin, Germany.

Yarkin, C., D. Sunding, D. Zilberman, and J. Siebert. 1994. Cancelling methyl bromide for postharvest use to trigger mixed economic results. California Agriculture, 48:16-21.

Zilberman, D., D. Sunding, M. Dobler, M. Campbell and A. Manale. 1994. Who makes pesticide use decisions: Implications for policymakers. In Pesticide Use and Product Quality, W. Armbruster (ed.). Glenbrook: Farm Foundation.

Index

T

U

V

W

Y